Geeks On Call® Wireless Networking

Geeks On Call® Wireless Networking

J. Geier
E. Geier
J. R. King

Wiley Publishing, Inc.

Geeks On Call® Wireless Networking: 5-Minute Fixes

Published by
Wiley Publishing, Inc.
10475 Crosspoint Boulevard
Indianapolis, IN 46256
www.wiley.com

Copyright © 2006 by Geeks On Call America, Inc., Norfolk, Virginia

Published by Wiley Publishing, Inc., Indianapolis, Indiana

Published simultaneously in Canada

ISBN-13: 978-0-471-77988-9
ISBN-10: 0-471-77988-1

Manufactured in the United States of America

10 9 8 7 6 5 4 3 2 1

1B/SY/RR/QV/IN

For general information on our other products and services or to obtain technical support, please contact our Customer Care Department within the U.S. at (800) 762-2974, outside the U.S. at (317) 572-3993 or fax (317) 572-4002.

Library of Congress Cataloging-in-Publication Data

Geier, Eric, 1984-
    Geeks On Call Wireless Networking : 5-Minute Fixes / Eric Geier, Jim Geier, J.R. King.
       p. cm.
    Includes index.
    ISBN-13: 978-0-471-77988-9 (pbk.)
    ISBN-10: 0-471-77988-1 (pbk.)
    1. Wireless LANs. I. Geier, James T. II. King, J. R., 1975- III. Title.
    TK5105.78.G45 2005
    004.6'8--dc22
                                                                              2005027876

Wiley also publishes its books in a variety of electronic formats. Some content that appears in print may not be available in electronic books.

Credits

Executive Editor
Carol Long

Development Editor
Sydney Jones

Production Editor
William A. Barton

Copy Editor
Kathryn Duggan

Editorial Manager
Mary Beth Wakefield

Production Manager
Tim Tate

Vice President and Executive Group Publisher
Richard Swadley

Vice President and Excecutive Publisher
Joseph B. Wikert

Project Coordinator
Ryan Steffen

Graphics and Production Specialists
Jennifer Heleine
Barbara Moore
Lynsey Osborn
Alicia B. South

Quality Control Technician
Brian H. Walls

Proofreading and Indexing
TECHBOOKS Production Services

Contents

Chapter 6: Security Settings   71

Part IV: Using and Maintaining Your Network 81

Chapter 7: Using Your Network   83

Chapter 8: Adding Peripherals to Your Network   97

Chapter 9: Maintaining Your Network   109

INTRODUCTION

Welcome to the Wonderful World of Wireless!

Wireless networks have been around for over a decade, providing mobility in warehouses, retail stores, and manufacturing plants. In recent years, the higher performance, improved security, and decreasing prices of wireless networks have made them practical and affordable for anyone to use at home or the office.

This book covers a wide range of topics about wireless networks, such as:

- **Know what you need before you buy:** In laymen's terms, we help you understand how wireless equipment works and what components you need for a wireless network.

- **Set up your physical network:** Learn how easy it is to connect your wireless network to the Internet.

- **Install and configure wireless components:** We show you how to install and properly configure wireless routers and wireless cards.

- **Protect your wireless network:** Because wireless networks use radio waves, you must use encryption and other security features to protect your data from high-tech criminals. We show you how easy it is to make your wireless network safe and secure.

- **How to use your wireless network:** We have lots of tips and tricks you can use to get the most out of your wireless network.

- **Add printers and other peripherals to your wireless network:** Take full advantage of your network by adding printers, video game adapters, and digital media players to your wireless world.

- **Maintain and troubleshoot your wireless network:** We show you how to keep your wireless network humming along by ensuring it has the latest updates. If anything goes wrong, we have tips about what to do.

PART I

WIRELESS NETWORKING BASICS

Before you go shopping for wireless equipment, you need to have an understanding of the basics. That way, you will know what components you need—which can save you time and prevent headaches when you actually install the network. This section of the book provides simple explanations about how wireless works and lists the items you need to set up a network.

1

Before You Buy

Before embarking on the installation of a wireless network, you should do your homework by thinking about what you need and the issues you might run into. Get off on the right foot by looking through this chapter and answering your initial questions before plopping down money on equipment.

There are many reasons why people use wired or wireless computer networks:

- **Share an Internet connection:** With a network, several people in your home or office can use the same high-speed Internet connection at the same time. For example, you may be researching your next vacation spot while your daughter is sending instant messages to her boyfriend.

- **Share files:** If you are tired of using floppies and USB thumbdrives to transfer files from one computer to the other, a network will make life much easier. After connecting to a network, you can simply drag and drop files from one computer to the other. After doing this a few times, you'll wonder how you ever made it through the day without a network.

- **Share a printer:** With a network, everyone can print to the same printer. For example, if you have a printer attached to a PC in your family room, you could print to it from your laptop while lounging on a hammock in the backyard. You can even attach a printer directly to the network, which will prevent someone who is printing a large document from slowing down the computer where the printer is attached.

- **Enjoy multiplayer games:** A network allows the use of interactive games, with each player sitting in front of his or her own computer. Many of the computer games on the market today have features that enable multiple players to take part in the action (assuming their computers can connect to a network). So, if you're into computer games and you have others who want to play, then don't wait. Install a network, now!

Why Go Wireless?

If you decide a network is right for you, then your next step is to select which kind you want: a traditional wired network (Ethernet) or a wireless network. Wireless, also called Wi-Fi, is rapidly growing in popularity. You have probably seen television commercials and other advertisements touting the benefits of wireless networks, or you may have friends and family who installed one. The following sections describe the advantages and disadvantages of wireless networks that you should consider when deciding if a wireless network is right for you.

Advantages of Wireless Networks

A wireless network provides the following advantages:

- **Mobility:** Similar to a cell phone, a wireless laptop or handheld computer (see Figure 1-1) enables you to communicate from just about anywhere. You're not forced to sit in front of a single desktop computer at a single location. Instead, you can use a wireless device to check your stocks while relaxing in front of the television (see Figure 1-2), check e-mail while cooking (see Figure 1-3), or find someplace quiet in the house to get some real work. In addition, you can take your wireless computer on the road and access the Internet from public Wi-Fi hotspots such as airports, hotels, universities, restaurants, and coffee shops.

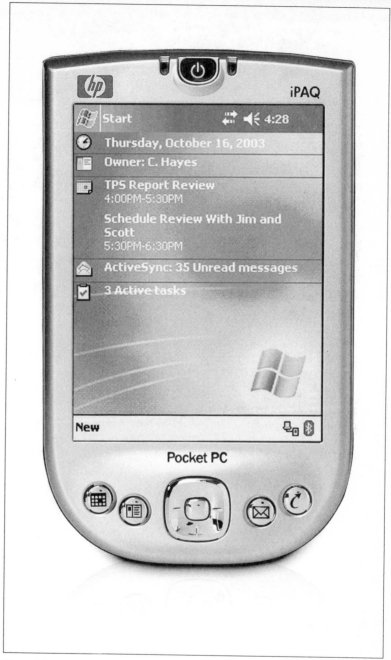

Figure 1-1

Figure 1-2

- **Easier installation:** A wireless network doesn't require you to run massive lengths of cable between two computers (which often requires you to spend hours fishing that cable between walls and ceilings).

- **Wide coverage area:** A wireless network's signal can cover a large amount of space. For example, if you have a photography studio in your garage, you can use a wireless network to connect to a computer in your house and share an Internet account.

Figure 1-3

Disadvantages of Wireless Networks

A wireless network can cause the following problems:

- **Interference:** Because a wireless network uses radio waves to send data between computers, other radio waves from microwave ovens and cordless phones may interfere with your network.

- **Security attacks:** Criminals can park outside a home or office and monitor or hijack the signals sent from wireless networks that have not been properly secured. In fact, there have been cases of thieves capturing wireless credit-card transactions by sitting in the parking lots outside stores. However, with the proper security techniques — which are available on all wireless network devices today — both you and the store stay safe. For information on encrypting and securing a wireless network, see Chapter 6.

- **The configuration is more complex:** Sometimes it can be tricky to properly configure a wireless router's security settings or other features. For that reason, pay close attention to Chapters 3 and 4.

FYI

Because wireless networks use radio waves, you may be concerned about health issues. Fortunately, research shows that the output power of wireless networks is much lower than cell phones, and there are no official reports of wireless networks causing any medical problems. However, as a precaution, you shouldn't touch the antenna of a wireless card or router while they are being used.

How Wireless Works

If you are installing your own wireless network, it is helpful to understand the big picture. Even if you're not technologically inclined, understanding the basic workings of your wireless network will provide you with some common sense in case you run into any problems. Here are some components and concepts you need to know:

- **Wireless card:** Each computer on the network must have a wireless card correctly installed and configured in order to send and receive wireless signals. These cards are easy to install and configure (as you will learn in Chapters 3 and 5).

- **Wireless router:** A wireless router is the main hardware in a wireless network. It links with a broadband modem to provide the network with a high-speed Internet connection. The router also sends radio signals that enable computers with wireless cards to connect to your network. You learn all about configuring a router in Chapter 4.

- **Medium access:** The wireless cards take turns sending data to and from each other over the air waves. Before a wireless card can transmit data, it must first analyze the air and determine whether another wireless card is transmitting a signal. If there is no signal present, then the wireless card can send data. If it detects a signal, the card waits and sends the data later. This "listen-before-transmit" method regulates access to the air and only allows one wireless card to send data at any given time.

- **Traffic flow:** In most networks, the digital traffic going from one wireless computer to another passes through a wireless router. For example, when Sierra sends a digital music file from her computer to Madison's computer, her computer transmits the file to the wireless router, and then the router sends it along to Madison's computer. Figure 1-4 shows a simple diagram of the way data flows across a network.

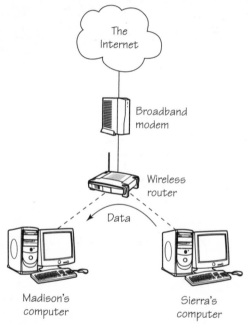

Figure 1-4

- **"Ad hoc" wireless network:** This type of network allows computers to communicate wirelessly with each other without using a router. You can swap files anywhere without having to connect to a wireless router (see Figure 1-5).

Figure 1-5

Got Wireless?

Most older computers do not have wireless cards, so those cards must be purchased separately. However, if your computer is relatively new, it might already have a wireless card installed. To find out via Windows or, if that doesn't work, you can look at your hardware, as described in the following sections.

Look Inside Windows

Follow these steps to use Windows to see if you have a wireless card:

1. Right-click the My Computer icon on your desktop. If this icon is not available, then click the Start button in the lower-left corner of Windows and click My Computer. If you can't find the My Computer icon anywhere, do the following:

    a. Right-click in the empty space on your desktop.

    b. Select Properties.

    c. A window opens. Click the Desktop tab.

    d. Near the bottom of the window, click the Customize Desktop button.

    e. Another window opens. On the General tab, beneath Desktop Icons, place a checkmark in the My Computer box.

    f. Click the OK button.

g. You will be returned to the previous screen. Click the Apply button.

h. Click the OK button.

i. The My Computer icon appears on your desktop. Right-click it.

2. Select Properties.

3. A window opens. Click the Hardware tab.

4. Click the Device Manager button.

5. Click the + (plus sign) located next to the Network Adapters category. If there are icons under Network Adapters, then you need to figure out what type they are. An Ethernet card (used with a traditional, wired network) has a label that says something like 10/100 or Ethernet. If the label mentions the word "wireless"—such as "wireless PCI card"—then it is a wireless card. Figure 1-6 shows a wireless card in the Device Manager.

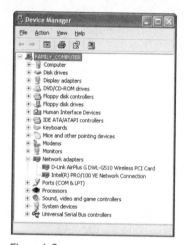

Figure 1-6

If you cannot find the Network Adapters category or if no icons appear under this category when you click the +, then most likely you do not have a wireless card installed. To be sure, you should look at your computer's hardware.

Look at Your Hardware

If you have looked in Windows but still can't determine whether you have a wireless card installed, then you should inspect your computer's hardware.

If you have a laptop computer:

- There is no quick way to determine whether you have an *internal* wireless card. Both the card and antenna are mounted inside your laptop behind an access door. In some cases, your laptop may have a manufacturer's label that indicates an internal wireless card exists.

• If you have an external wireless card, it sticks out of a slot on the side of your laptop (see Figure 1-7).

Figure 1-7

Follow these steps if you have a desktop computer:

1. Move your computer's case so you can see its back. Be careful not to yank out or disconnect any wires.

2. If you have a wireless card installed, an antenna several inches long extends from the back of the computer. This antenna is attached to a card mounted inside the computer's case (see Figure 1-8).

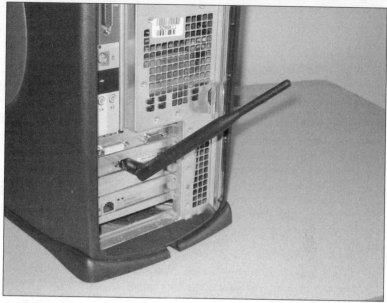

Figure 1-8

3. Some wireless cards have desktop antennas connected to them. You can check for this by following any wires coming from your computer's cards.

The Lowdown on Wi-Fi

Wireless Fidelity (Wi-Fi) is a brand name given to wireless networks by the Wi-Fi Alliance (an organization that governs Wi-Fi products). There are several versions of Wi-Fi: 802.11a, 802.11b, and 802.11g, each of which uses different methods for transmitting and receiving data. For more information about their differences, see Table 1-1.

Table 1-1: Wi-Fi Standards

Technology	Frequency Band	Speed	Uses
802.11b	2.4 GHz	Up to 11 Mbps	Ideal for older wireless networks
802.11g	2.4 GHz	Up to 54 Mbps	Great for e-mail and Web browsing
802.11g with SpeedBooster	2.4 GHz	Up to 108 Mbps	Enables faster file downloads
802.11a	5 Ghz	Up to 54 Mbps	Delivers higher performance when significant radio-frequency interference is present

It is important to understand that the 802.11b and 802.11g technologies are compatible with each other, so even if you have older 802.11b equipment, it still works with the newer 802.11g equipment. Also, be aware that in order to use the 802.11g "SpeedBooster" feature, both your wireless router and wireless card must support it.

Ensure Compatibility

Because there are several different standards of wireless networks, you must buy a wireless card that is compatible with your wireless router. Otherwise, the devices won't be able to "talk" to each other. In addition, if you are planning to use special features like SpeedBooster or power saving, you must make sure those features are properly set up. Here are some general guidelines for ensuring compatibility:

- **Use compatible technologies:** Make sure the wireless card and wireless router are using the same frequency. For example, a 2.4 GHz wireless card (802.11b or 802.11g) can only connect to a 2.4 GHz wireless router (802.11b or 802.11g). It cannot connect to a 5 GHz (802.11a) wireless router.

- **Buy Wi-Fi certified hardware:** Look for the Wi-Fi certification listed on the product packaging or on the product itself. You can also visit the Wi-Fi Alliance's website (www.wi-fi.com) to view the products that have Wi-Fi certification.

- **Use the same manufacturer:** If possible, use the same brand of wireless cards and routers. This allows you to take advantage of special manufacturer-specific enhancements, such as SpeedBooster or range-extension techniques.

It's a Fact

The number 802.11 comes from the group that developed the initial standards for local area networks. The 80 refers to the year they began their work—1980—and the 2 refers to the actual month they began—February, the second month. The 11 was later given to the 802 group that created the wireless LAN standard because it was the eleventh standard they had developed.

FYI

If you want wireless equipment that can work with any 802.11 standard (a, b, or g), buy a "dual-band" router and card.

Increase the Lifespan of Your Wireless Equipment

As with any other computer technology, wireless networks have rapidly evolved, becoming faster and more secure. Newer wireless equipment should last for many years until something better comes along that is more appealing. Here are a few tips to keep your wireless equipment from becoming outdated too soon:

- **Buy Wi-Fi certified equipment:** This can't be said enough. If you don't purchase devices that have an official Wi-Fi certification, you run the risk of having that equipment become troublesome or unusable in the near future.

- **Check for firmware and driver updates:** Periodically, hardware manufacturers release two kinds of updates for their equipment: firmware and drivers. These updates fix problems, offer enhanced features, and plug security holes. It is a good idea to check for new firmware and drivers every few months just to stay safe. Some manufacturers send e-mail announcements to inform you when updates are available, but others require you to check their website.

The Equipment You'll Need for a Network

To set up a wireless network, you need to learn some high-tech terminology and purchase special equipment.

Untangle the Terminology

Whether you're browsing the shelves in a computer store or researching Wi-Fi products on the Internet, you might be confused by the high-tech lingo. Often there are several names that mean the same thing, so here is a quick rundown of common Wi-Fi words and their synonyms:

- **Wireless router:** Also called a wireless broadband router or an access point.

- **Wireless card:** Also called a PCI adapter, desktop adapter, Wi-Fi adapter, or wireless adapter.

- **Notebook adapter:** Also called a cardbus adapter, wireless PC card, radio card, Wi-Fi adapter, or wireless adapter.

- **2.4 GHz:** Also called 802.11b, 802.11g, b, or g.

- **5 GHz:** Also called 802.11a, Wi-Fi5, or a.

Wireless Router

The average wireless router (as shown in Figure 1-9) has a range of 100 feet, so if you are interested in setting up a wireless network in your home or small office, you should only need one router. Some routers are "single-band" and implement either 2.4 GHz technologies (802.11b and 802.11g) or 5 GHz technologies (802.11a). Some routers, though, are "dual-band," which means they use both 2.4 GHz and 5 GHz frequencies. Consider the following when deciding whether to purchase a dual-band router:

- **Price:** The price of dual-band routers is generally 25 percent higher than a standalone 2.4 GHz (802.11g) router.

- **Compatibility:** Because the majority of laptops and PCs come equipped with a 2.4 GHz (802.11b or 802.11g) wireless card, you can probably get by with only having an 802.11g router. Don't forget, a 2.4 GHz wireless card can't connect to a 5 GHz router, so no matter what type you choose, always use wireless equipment that shares the same frequency.

• **Performance:** There is much less interference in the 5 GHz band, which means that 802.11a routers can operate with fewer interruptions than 802.11g routers.

Figure 1-9

Wireless Card

There are several types of wireless cards to choose from:

• **PCI card:** You'll need a wireless PCI card for each computer you want to connect to a wireless network. The card shown in Figure 1-10 inserts into your computer.

Figure 1-10

- **Notebook adapters:** You need a wireless notebook card (see Figure 1-11) for each laptop or handheld computer. Before spending money on equipment, make sure your handheld computer has a notebook adapter slot. Usually this slot is located on the top of the computer. Also, look in your owner's manual to verify whether your handheld computer will accommodate a notebook adapter.

Figure 1-11

- **Compact Flash adapter:** You may need a wireless Compact Flash (CF) adapter for each handheld computer you want to connect to a wireless network (see Figure 1-12). Before spending money on equipment, make sure your handheld computer has a CF adapter slot. Usually this slot is located on the top of the computer. Also, look in your owner's manual to verify whether your handheld computer will accommodate a CF card.

Figure 1-12

- **USB adapters:** Use this type if you have an available USB port and want to avoid the hassles of opening up your desktop computer and inserting a card.

- **Antenna:** In most cases you can use the antenna that came with your wireless card. However, if the antenna is removable, you can purchase a replacement that is more powerful.

Determine the Number of Cards You Need

Here are some things to consider when deciding how many wireless cards you should purchase:

- Buy one wireless card for each computer you want to connect to the network.

- You can save money by using an inexpensive Ethernet cable to connect your desktop or laptop computer to the traditional wired ports on the back of the wireless router.

Add Bells and Whistles

To boost the performance of your wireless network, there are a few enhancements you can use, as described in the following sections.

MIMO

MIMO (which stands for multiple input, multiple output) is a feature appearing in new routers that enables you to improve the performance of your wireless signal by using "smart" antennas. Here are some things to consider about MIMO:

- **Standards:** Currently there are no official standards for MIMO. It is being incorporated into a new wireless standard called 802.11n, but the approval process has been slow. As a result, the standard may not become available until 2007. Some manufacturers have already produced MIMO-based wireless routers, but these are unofficial.

- **Do you need MIMO?** The answer is probably no, especially if your home or office can get by with a single router without MIMO. Routers with MIMO cost more, so if you don't need it, don't buy it.

SpeedBooster

Some wireless hardware has an enhancement known as "SpeedBooster" that doubles its speed. For example, it can increase the data rate of an 802.11g router from 54 Mbps to 108 Mbps. If you purchase a wireless product with this feature, consider the following tips:

- **Use the same brand of equipment:** Because SpeedBooster is not an official Wi-Fi standard, you have to use wireless cards and routers from the same manufacturer.

- **Enable it on all equipment:** If you want to take advantage of SpeedBooster, you must turn on this feature in the wireless router and in each wireless card.

Upgrade the Antenna

In most cases, you can use the antenna that came with your wireless router. However, if the antenna is removable, you can replace it with one that is stronger and has a better range. Here are some general guidelines:

- **Visit the manufacturer's website:** By researching the styles of antennas offered by the manufacturer of your wireless router, you will be able to buy one that properly fits your router. Don't forget: you can't replace an antenna that isn't removable.

- **Purchase from a third party:** You can purchase a variety of antennas for Wi-Fi devices from companies like RF Linx (www.rflinx.com) and Til Tek (www.tiltek.com).

- **Check for compatibility:** When you order antennas, be sure they are designed to work with your brand of wireless equipment. Different manufacturers use different types of antenna connectors, so it is important to get this right. Otherwise, you won't be able to attach the new antenna to your device.

Hardware and Software Requirements

Wireless cards are designed to work with a variety of operating systems, such as Windows, Mac OS, and Linux. The memory and processor speeds are minimal, so it is unlikely you will encounter any issues with compatibility. Here are some tips to consider:

- **Use Windows XP:** The most recent version of Windows, known as Windows XP, makes it easy to set up and use wireless networks. If you don't have XP, you can buy a copy and perform a manual upgrade or you can have Geeks On Call come to your home or office to do the upgrade.

- **Have a CD drive available:** Make sure your computer has a CD-ROM drive, because you probably need to install software for your wireless router or card that comes on a CD-ROM.

Estimate Your Network's Cost

Here are some approximate prices for the devices you must purchase to set up a wireless network:

- **Wireless adapter:** $20 to $80
- **Wireless router:** $30 to $100

Which Manufacturer Is Best?

There are several major companies that produce outstanding wireless network products for homes and small businesses. Here are some respected names (in no particular order):

- Buffalo: www.buffalotech.com
- Linksys: www.linksys.com
- D-Link: www.dlink.com
- Netgear: www.netgear.com

FYI

If you already know what wireless products you want to buy, consider purchasing them from an Internet store like Newegg (www.newegg.com). Often this can save you quite a bit of money.

To pick the best brand of wireless products for your needs, consider the following:

- **Price.** When shopping around to get the lowest possible price on wireless equipment, keep in mind that cheaper is not always better. Don't go with no-name brands just to save a few bucks. Stick with a reputable manufacturer's products.

- **Enhancements.** Look for special features like MIMO and SpeedBooster. Investigate whether the manufacturer also sells wireless companion products such as digital media players, cameras, and print servers. If so, your network has room to grow and won't have problems connecting to those other products.

- **User friendliness.** Look at the packaging to see if there are any clues about how easy the product will be to install. However, don't base your entire decision on the package.

The Effects of Bad Weather

Heavy rain, fog, and other undesirable weather conditions decrease the range of some wireless networks, such as those providing wireless connections between buildings in a city. But for small wireless networks less than 500 feet in diameter— like the kind found in the average home or office—the impact of bad weather is undetectable.

How Many Computers Does a Network Support?

The average wireless router can support up to 20 wireless computers. However, using an Internet telephone like Vonage (referred to as voice over IP, or VoIP for short) may strain the network and make it feel sluggish. If this happens, consider using two or more routers in the same area, but set them to different channels.

PART II

INSTALLING YOUR ROUTER AND WIRELESS CARDS

Once you have purchased the equipment for your wireless network, it's time to install it. You will need to insert a wireless card in your computer and connect a wireless router to your high-speed Internet service. Since wireless networks don't have many cables, the installation of the network is very easy. Just follow these simple tips and you will be surfing the wireless world in no time.

2

Setting Up Your Physical Network

The purchase, installation, and configuration of your new wireless network can take several hours. Before investing all of that time (and money), read through this general overview of the steps involved:

1. Find a location in your home or office to set up the wireless router. This may require some trial and error to find the optimum spot.

2. Install and configure your wireless router.

3. Install and configure a wireless card in each computer.

4. After you finish the installation, check to make sure that each computer can connect to the router and can browse the Internet.

5. Install and configure any optional wireless components. For example, you may want to install a wireless print server, video game adapter, digital media player, or webcam.

Find the Best Place to Install Your Router

One of the most important steps in setting up a wireless network is to place the router in a location that gives you the best possible wireless access and coverage. Before picking a spot, think about the areas in your home or office where you will frequently use your wireless or wired computers and mobile devices. Here are some tips for finding a good location:

- **Install near the broadband modem:** If you are planning to share an Internet connection with two or more computers, then you should install your wireless router near the DSL, satellite, or cable modem. Doing so will allow you to easily run a

cable between the modem and the router. If this location doesn't provide adequate wireless coverage, you can move your modem to a different wall outlet (in which case you may need the services of a
computer professional like Geeks On Call).

- **Install near the center of your coverage area:** After deciding the
  general area of your home or office where you want to have a wireless
  network, you should install the wireless router in the center of that
  area. This will ensure that your wireless computers and mobile
  devices have access to strong wireless signals no matter where you go
  in your coverage area.

Types of Wireless Networks

There are three types of wireless networks: a wireless router connected to high-
speed Internet, an ad hoc network, and a partially wired network.

Wireless Router Connected to High-Speed Internet

The most common type of wireless network uses a wireless router to regulate
the flow of information and connect you to a traditional wired network (like the
Internet). See Figure 2-1 for a diagram of a network that has a router connected
to high-speed Internet.

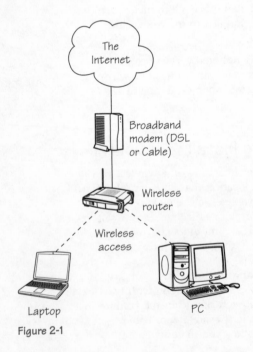

Figure 2-1

Here are common reasons why you might choose this type of network:

- You need to connect your computer to the Internet.
- You want to share an Internet connection with multiple computers.
- You want to access public wireless networks.

Connect the Router to a Broadband Modem

Here is a general overview of the steps required to connect a router to a high-speed Internet connection (also known as a broadband connection):

1. Sign up for Internet access with a broadband service provider, such as cable, satellite, or DSL. Your service company installs a broadband outlet in your home or office and likely connects a modem (see Figure 2-2) to this outlet. If you don't receive a broadband modem as part of your subscription, you need to rent or purchase one. Check with your service provider for information on obtaining a compatible modem.

Figure 2-2

2. If your service provider didn't connect the modem, then you need to connect it to the wall outlet by using the Category 5 Ethernet cable (often called a CAT5 cable) supplied by your provider. Generally they are gray, yellow, or blue in color and resemble a large telephone cord (see Figure 2-3). If a CAT5 cable didn't come bundled with your router or modem, then you must purchase one from a computer or electronics store.

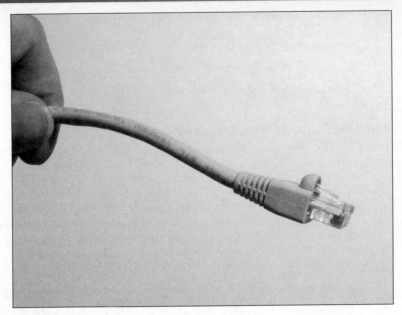

Figure 2-3

3. Plug one end of the CAT5 cable into the slot on the router labeled "WAN" or "Internet." You will know the cable has been inserted properly when you hear it snap into place.

4. Plug the other end of the CAT5 cable into the slot on the broadband modem labeled "LAN" or "Network." You will know the cable has been inserted properly when you hear it snap into place.

FYI

The slots on the back of some broadband modems do not have labels like WAN or Internet. In that case, find the slot that looks like it will connect to a large telephone cord. That is where you should plug your CAT5 cable.

An Ad Hoc Network

Another type of wireless network known as "ad hoc" (also called "peer-to-peer") allows computers to communicate wirelessly with each other without using a router (see Figure 2-4).

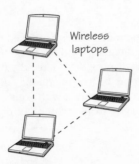
Wireless laptops

Figure 2-4

You might consider using this type of network if . . .

- There aren't any wireless routers installed at your location but you want to transfer a file from one wireless computer to another.

- You don't need to connect to the Internet.

- You are away from your home or office but want to use a network to swap files, share printers, or play multiplayer video games.

The following sections describe the methods for setting up an ad hoc network.

Use the Manufacturer's Configuration Utility

The most common way to set up an ad hoc network is to use the special software that comes with your wireless card. This software is usually referred to as a "configuration utility." To use a configuration utility, follow these steps:

1. Access your wireless card's configuration utility.

   a. Usually this can be done by double-clicking the configuration utility's icon located in the lower-right corner of Windows.

   b. If you don't see it, click the Start menu in the lower-left corner of Windows, click All Programs, and then select the configuration utility that corresponds to the name of your wireless card.

2. Once you have successfully accessed the configuration utility, look for the SSID feature. This is the name of your ad hoc network. You should change the name to one of your choosing (feel free to be creative).

Note

For enhanced security, use a combination of letters and numbers. Also, don't forget that an SSID is case sensitive, which means it considers uppercase and lowercase letters to be different.

3. Choose a channel for your ad hoc network. Usually it doesn't matter which channel you select, but try to pick one that is different from those being used by nearby routers (this will reduce the possibility of interference from the other routers).

4. Change the wireless mode to Ad Hoc (or, in some cases, you may have to select Peer-to-Peer).

5. Click Apply to save the changes.

Use Windows XP's Configuration Utility

Another way to set up an ad hoc network is to use the built-in features of Windows XP, as follows:

1. Right-click the wireless icon located in the lower-right corner of Windows (see Figure 2-5).

Figure 2-5

2. Click View Available Wireless Networks.

3. If your version of Windows XP has Service Pack 2 installed, click the Change Advanced Settings option. If you don't have Service Pack 2, click the Advanced option (see Figure 2-6).

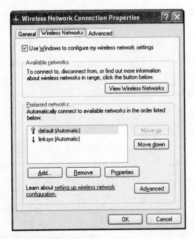

Figure 2-6

4. Click the Wireless Networks tab.

5. Click Add.

6. In the Network Name (SSID box), type a new name for your ad hoc network. You can create any name you want, so feel free to be creative.

Note

For enhanced security, use a combination of letters and numbers. Also, don't forget that an SSID is case sensitive, which means it considers uppercase and lowercase letters to be different.

7. Near the bottom of the window, put a check mark in the box labeled "This is a computer-to-computer (ad-hoc) network; wireless access points are not used" (see Figure 2-7).

Figure 2-7

8. To protect your ad hoc network from hackers, you should turn on its encryption features, as follows:

    a. Click the Data Encryption drop-down menu.

    b. Select an encryption type.

    c. Type a password.

9. Click the OK button. The wireless card starts operating as an ad hoc device, meaning it will not connect to any routers. As a result, you cannot use the Internet unless you set up Internet-sharing on one of the computers.

Note
Other people can easily connect their computers to your ad hoc network. First, they must access their wireless card's configuration utility, select the Display option, and connect to ad hoc networks. Next, they should right-click the wireless icon located in the lower-right corner of Windows and select the View Available Wireless Networks option. Finally, they must select the SSID name you created for your ad hoc network.

A Partially Wired Network

On the back of most wireless routers are Ethernet connections identical to those found on a traditional wired router. This allows the wireless router to share files and an Internet connection with computers that don't have wireless cards. This setup is known as a "partially wired network" (see Figure 2-8).

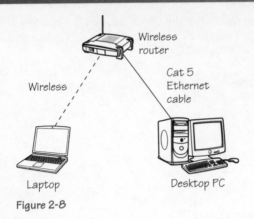

Figure 2-8

Note

If there aren't enough ports available on the wireless router, you can connect the router to an Ethernet hub.

Set Up a Partially Wired Network

Assembling a partially wired network is easier than it may seem. Just follow these steps:

1. Make sure your computer has an Ethernet card. To do so, look for an opening in the back of the computer that seems like it could connect to a large telephone cord. This opening might be labeled "Data," "Ethernet," or "RJ-45."

> **FYI**
> The CAT5 cable must be less than 300 feet long.

2. Obtain a CAT5 cable long enough to reach from the computer to the wireless router.

3. On the back of most routers are Ethernet ports numbered from 1 through 4. Plug the CAT5 cable into any of these ports by snapping the connector into place (see Figure 2-9).

4. Connect the other end of the cable to one of the Ethernet ports on the wireless router.

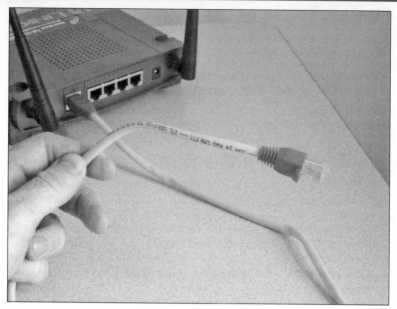

Figure 2-9

Connect a Wireless Router to an Existing Wired Network

Even if you already have a traditional wired network, you can easily add a wireless network. Doing so will enable you to share files and an Internet connection with mobile computers or other mobile devices you may purchase in the near future (see Figure 2-10).

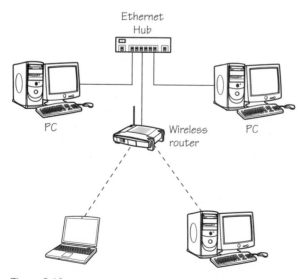

Figure 2-10

1. Plug a CAT5 cable into the LAN port on your wireless router.

2. Connect the other end of the cable to a hub or wired router on your existing wired network.

3. Install and configure the wireless router, following the manufacturer's installation guide. For more information on configuring routers, see Chapter 3.

High-Tech Fun in the Sun

Have you ever thought about how much fun it would be to surf the Web while relaxing in a hammock in your backyard or lounging by your pool? With wireless networks, it's possible. If you have tried to extend the range of your indoor wireless network but it won't reach outdoors, you will probably need to install a router outside. This can take some effort, so be sure you really want wireless coverage outside before going any further. Figure 2-11 shows you how an exterior router can connect to a wireless network you already have installed inside your home or office.

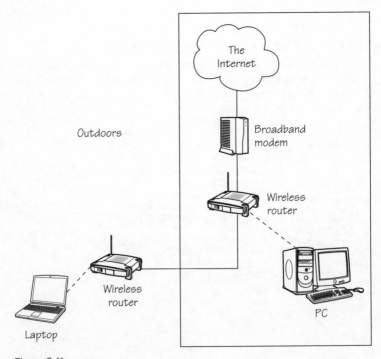

Figure 2-11

Install a Wireless Router Outdoors

Here are the basic steps you must take to install a wireless router outside:

1. Obtain a wireless router rated for outdoor installations, or use a regular interior wireless router in combination with a special weatherproof enclosure recommended by the router's manufacturer. Usually these enclosures aren't available at the average computer or electronics store, so you will probably need to purchase one on the Internet, either from the manufacturer's website or from an Internet store.

2. Find a good exterior location that will prevent large amounts rain and snow from hitting the router, such as underneath the edge of your roof.

3. Mount the router. Most exterior routers come with plenty of mounting brackets.

4. Connect the exterior router to your existing interior wireless network, as follows:

   a. Purchase a CAT5 cable long enough to stretch from the exterior router to the interior router.

   b. Run the cable from the exterior router to the interior router. This may require you to drill holes in the side of your house and fish the cable through walls.

   c. Plug one end of the cable into an available Ethernet port on the exterior router, and then plug the other end of the cable into an available Ethernet port on the interior router.

   d. Power up the exterior router. If the Ethernet cable doesn't supply electricity to the router (which is called "power-over-Ethernet"), then use an extension cord to plug the exterior router into an exterior electrical outlet.

5. Configure the router (see Chapter 4 for more details). Be sure to do the following:

   a. Change the name of your exterior router (its SSID) to match the SSID of the interior router. This allows you to roam from inside to outside without having to reconnect to both networks each time.

   b. Change the channel of the exterior router so it does not use the same one as the interior router. To avoid interference, only use channels 1, 6, and 11.

   c. Configure the exterior router to give it the same security settings as your interior router.

   d. To avoid conflicts, disable the "DHCP" feature on the exterior router.

3

Installing a
Wireless Card

T his chapter guides you through the process of installing a
wireless card that enables your computer to "talk" to the
wireless router. Figure 3-1 shows a typical wireless card.

Figure 3-1

Do It Yourself

Consult the user
guide

Install a wireless card
in a desktop
computer

Install a wireless card
in a laptop

Window XP's Wireless
Auto Configuration

Consult the User Guide

Your product's user guide contains answers to frequently answered questions, valuable information about how to access configuration screens and trouble-shoot problems, and much more. When you are installing a wireless network or attempting to solve a problem with it, read the user guide first.

Most wireless network products don't come with a printed user guide, so you need to locate it on the CD-ROM that came with your hardware. In many cases, the user guide can be accessed through a window that pops up after you insert the CD. If no pop-up window appears, you can manually search for the user guide by doing the following:

1. Double-click the My Computer icon on your desktop. If this icon is not available, then click the Start button in the lower-left corner of Windows and click My Computer. If you can't find the My Computer icon anywhere, do the following:

   a. Right-click in the empty space on your desktop.

   b. Select Properties.

   c. A window opens. Click the Desktop tab.

   d. Click the Customize Desktop button near the bottom of the window.

   e. Another window opens. On the General tab, beneath Desktop Icons, place a check mark in the My Computer box.

   f. Click the OK button.

   g. You are returned to the previous screen. Click the Apply button.

   h. Click the OK button.

   i. The My Computer icon appears on your desk-top. Double-click it.

2. Right-click the icon for your CD or DVD drive, and then click Explore.

3. The user guide is often found in a folder on the Docs or Manual CD. The name of the user guide usually contains the model number for your product.

FYI
User guides typically come in PDF format, which requires the Adobe Acrobat Reader to view them. If this software isn't installed on your computer, check the setup CD that came with your wireless product to see if a copy is included. If not, you can download Acrobat Reader free of charge at http://www.adobe.com.

4. If you have lost (or "temporarily misplaced") your product's CD, you can usually download the user guide from the Support section of the manufacturer's website. Make sure you know the model number of your product in order to download the correct manual.

Install a Wireless Card in a Desktop Computer

When installing a wireless card in a desktop computer, always follow the instructions that came with the card. Here is a general overview of the steps involved:

1. Follow the proper instructions: Most user guides include different instructions for different versions of Windows (such as Windows XP or Windows 98), so be sure to follow the set of instructions that were designed for your version of Windows.

2. Turn off the computer and unplug it: To protect yourself from being shocked or damaging the computer's components, you must shut down Windows, turn off the computer, and unplug it from all electrical outlets.

3. Open the computer's case: Almost every desktop computer allows you to open its case for the purpose of accessing or changing its components. Depending on how your desktop computer was built, the whole case may slide open, one side may slide open, or one side may swing open like a door.

4. Insert the wireless card: After opening the computer's case, look for an empty PCI slot on the motherboard where you can insert the wireless card. See Figure 3-2 to get an idea of what a PCI slot looks like. Before inserting the card, you will probably need to remove a small metal cover located adjacent to the PCI slot. This can be done by removing the screw that holds the cover in place. Again, refer to Figure 3-2 for an example of what the slot looks like with the cover removed. When sliding the wireless card into a PCI slot, make sure its snaps into place and is not loose. The side the card with the antenna should stick out the back of the computer through the open slot.

5. Install the software and configure the card: Sometimes Windows detects and configures a wireless card without asking for additional software, but it is always a good idea to follow the instructions provided by the card's manufacturer and install the special software that came with the card. Typically this software includes updates and utilities that make the installation easier and enhance the card's performance. For more information on installing this software and configuring a wireless card, see Chapter 5.

FYI
Read the manufacturer's instructions carefully because you may be asked to install the software before physically installing the wireless card (or vice versa).

Figure 3-2

Install a Wireless Card in a Laptop

When installing a wireless card in a laptop computer, always follow the instructions that came with the card. Here is a general overview of the steps involved:

1. **Obtain a notebook wireless card:** This is the only type of wireless card that will plug into your laptop. You don't have to buy the card from the same manufacturer as your wireless router, but make sure the card and the router use the same frequency (either 2.4 GHz or 5 GHz).

2. **Connect a CD-ROM drive (if necessary):** If the installation requires a CD-ROM to be used (which is likely), make sure you have a CD drive. Some lightweight laptops don't come with a CD drive, so you might need to use an external drive. If you don't have any type of CD drive available, you might be able to download the software for your wireless card by visiting the manufacturer's website.

3. Follow the proper instructions: Most user guides include different instructions for different versions of Windows (such as Windows XP or Windows 98), so be sure to follow the set of instructions that were designed for your version of Windows.

4. Insert the wireless card: Choose an empty slot on your laptop (you probably only have one), and then insert the card (see Figure 3-3). Although the card can only go in one way, be gentle because the card and the slot are somewhat fragile.

Figure 3-3

FYI

Read the manufacturer's instructions carefully because you may be asked to install the software before physically installing the wireless card (or vice versa).

5. Install the software and configure the card: Sometimes Windows detects and configures a wireless card without asking for additional software, but it is always a good idea to follow the instructions provided by the card's manufacturer and install the special software that came with the card. Typically this software includes updates and utilities that make the installation easier and enhance the card's performance. For more information on installing this software and configuring a wireless card, see Chapter 5.

Update the Drivers for Your Wireless Card

To get the best performance from your wireless card, you should update it with the most recent "drivers," which are a type of software that controls your card. Occasionally manufacturers release new drivers for their products to offer enhanced features, fix errors, and plug security holes.

Determine the Current Version

Before making changes to your drivers, you must find out what version they are.

Follow these steps to determine the version of your wireless card drivers on Windows XP Home Edition and XP Professional Edition:

1. Double-click the wireless icon located in the lower-right corner of Windows.

2. Select Properties.

3. Click Configure.

4. Click the Driver tab (see Figure 3-4).

Figure 3-4

5. Grab a pen and paper and jot down the version number and release date for your wireless card's drivers.

Follow these steps to determine the version of your wireless card drivers on Windows 98:

1. Right-click the My Computer icon on your desktop.

2. Select Properties.

3. Click the Device Manager tab.

4. Click the + (plus sign) located next to the Network Adapters category.

5. Double-click the wireless card you want to check.

6. Click the Driver tab.

7. Click Driver File Details.

8. Grab a pen and paper and jot down the version number and release date for your wireless card's drivers.

Check for New Drivers

To determine if your drivers need to be updated, do the following:

1. Visit the website belonging to the manufacturer of your wireless card. Click the Support or Downloads section of the site either, and then look for information about drivers.

2. Locate the drivers on the website that correspond to your specific brand and model of wireless card.

3. Compare the version number of your current drivers with the ones available on the website. If the website's drivers have a higher number than yours, that means your current drivers are outdated. Go ahead and download the new ones. Be sure to save them in a folder you can easily find later, because you will need to access that folder when you update the drivers in the next step.

Update the Drivers

After downloading the new drivers, you need to update your wireless card through Windows.

To update your wireless card on Windows XP Home Edition and XP Professional Edition:

1. Double-click the wireless icon located in the lower-right corner of Windows.

2. Select Properties.

3. Click Configure.

4. Click the Driver tab.

5. Click Update Driver (see Figure 3-5).

Figure 3-5

6. Follow the on-screen instructions.

To update your wireless card on Windows 98:

1. Right-click the My Computer icon on your desktop.

2. Select Properties.

3. Click the Device Manager tab.

4. Click the + (plus sign) located next to the Network Adapters category.

5. Double-click the wireless card you want to update.

6. Click the Driver tab.

7. Click Update Driver.

8. Follow the on-screen instructions.

Windows XP's Wireless Auto Configuration

Windows XP offers Zero Configuration and Wireless Auto Configuration services that make it easier to use wireless networks. With Wireless Auto Configuration, you don't have to fiddle with the often confusing configuration utilities that come with wireless cards. Here is how the Wireless Auto Configuration works:

1. The Wireless Auto Configuration prompts you with a message that says that wireless networks have been detected (see Figure 3-6).

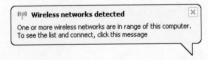

Figure 3-6

2. Open the Wireless Network Connection dialog box.

3. Select the network you want to connect to (see Figure 3-7).

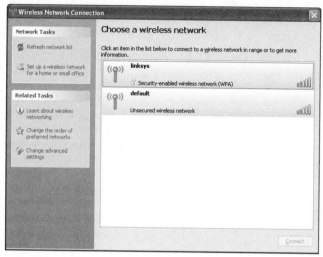

Figure 3-7

4. Windows attempts to connect to the network. If the network is protected with security features, you may be required to enter a WEP key or WPA key (which are similar to passwords).

One of the following occurs:

- If Windows successfully connects to the network, that network is automatically added to your list of preferred networks (see Figure 3-8).

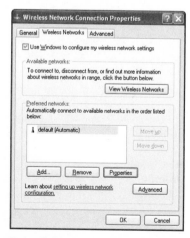

Figure 3-8

- If Windows can't connect to any of the networks that are currently available, it tries to connect to the ones found in your list of preferred networks (previously shown in Figure 3-7). If Windows can't connect to those preferred networks, it uses your wireless card to scan for the presence of any wireless networks within range.

- If you are using the Windows "Automatically connect to non-preferred networks" feature (see Figure 3-9), your computer automatically connects to the first wireless network it finds. If you have disabled this feature (as in Figure 3-10), Windows does not connect to any wireless networks it finds. Instead, Windows displays a message that says, "One or more wireless networks are available" (previously shown in Figure 3-6). This gives you complete control over which wireless network your computer accesses.

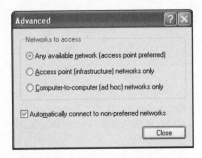

Figure 3-9

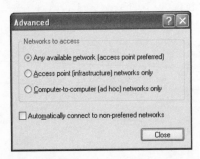

Figure 3-10

PART III

CONFIGURING YOUR ROUTER AND WIRELESS CARDS

To get maximum performance and security out of your wireless network, you need to spend some time configuring and tweaking your wireless router and wireless cards. Some of these tasks should be done during the installation of your network, while others can be done after you have spent a little time using your wireless network. This section of the book uses simple terms to describe the changes to make and the optimal settings to use.

4

Configuring Your Router

5-Minute Fixes

A wireless router is the most important piece of hardware in a wireless network, which means that learning how to configure the router is the most important part of setting up a network. Here is a general overview of the critical features you should change during the initial setup:

- **Password for the configuration utility:** Most wireless routers use common, default passwords that are well known to criminals. If you don't change this default password, an intruder could easily access your router's configuration utility and change your settings.

- **Network name (SSID):** At the time of their manufacturing, routers belonging to the same brand are given identical names (SSIDs). You should change your router's SSID to increase your security and prevent conflicts with nearby routers made by the same company.

- **Wireless channel:** By default, many wireless routers have their channels set to 6. Although this should work okay, you might have some conflicts with neighboring routers that are using the same channel. To avoid potential problems, change your channel to 1 or 11.

- **Encryption method:** To prevent criminals from hijacking your wireless signals and viewing your private e-mail, passwords, and bank account information, you must use one of the security modes on your wireless router and wireless cards. For networks in a home or small office, use the WPA-PSK security feature. If that's not available, use WEP.

Do It Yourself

Configure your router for the first time

Set a static IP for your router if necessary

Enable DHCP

Set the SSID for your router

Disable SSID broadcasting

Change the default password

Configure advanced settings

Configure Your Router for the First Time

Although your wireless router will probably work just fine using its default settings, you should configure it to boost its performance and increase its security. Before attempting to do so, carefully read the instructions that came with the router.

Understand IP Addresses

An Internet Protocol (IP) address is a number given to each device that connects to a wired or wireless network. Think of it like a mailing address: just like you need to know someone's address to mail him or her a package, computer networks need IP addresses in order to deliver data to the proper destination. Before your computer can connect to the Internet, it must have an IP address. When you set up a wireless network, your wireless router will automatically assign an IP address to your wireless card. However, in some cases an Internet service provider might require you to manually set the IP address in your router. These IP addresses are referred to as "static" addresses because they stay the same and do not change automatically.

Find Your Router's IP Address

To change or tweak the features in your wireless router, you are required to use a special configuration utility that can only be accessed by typing the router's IP address into your Web browser (like Internet Explorer). If you don't know what your router's IP address is, try one of the 5-minute fixes provided here.

Try a Common Address

The first thing you should try is one of the common, default IP addresses used by many routers:

- 192.168.1.1
- 192.168.0.1
- 192.168.2.1

Consult Your Manual

Most user guides, manuals, or instructions that come with a router indicate what IP address should be used. If you don't have these documents or if they don't mention anything about IP addresses, follow one of the other 5-minute fixes.

FYI
Some wireless routers may require you to login through a website address like www.router login.net rather than using an IP address.

Look in Windows

If none of the commonly-used IP addresses work with your router, or if your user guide isn't helpful, you can determine your IP address by looking inside Windows. This requires you to be connected to your wireless network. Follow these steps to look up the address:

1. Double-click the wireless icon located in the lower-right corner of Windows. The wireless connection status window opens.

2. Click the Support tab.

3. View the Default Gateway value. This is the IP address of your router.

Set a Static IP for Your Router If Necessary

When you are setting up your wireless router, contact your Internet provider and ask them if they require a static, unchanging IP address. If so, you will have to enter it into your router before you can connect to the Internet. If you aren't required to use a static IP address, make sure your router is using a feature known as "DHCP." Fortunately, most routers have this turned on by default. For more information on DHCP, see the 5-minute fix in this chapter titled "Enable DHCP."

Follow these steps to set up a static IP address:

1. Log in to your wireless router's configuration utility, as follows:

   a. Make sure you are connected to your wireless network. Test this by hovering your mouse pointer over the wireless icon located in the lower-right corner of Windows. A message should indicate your connected status.

   b. Open a Web browser like Internet Explorer. If your router is not connected to the Internet, your browser will not find a valid Web page. Instead, it displays an error message that says something like "Page cannot be displayed" or "Page not found."

   c. In your Web browser's address window (the place where you usually type the name of a website you want to visit), type your router's IP address then press the Enter key. Do not type http:// or www. Instead, only type the number for your IP address.

   d. A login box appears. If you created a username and password when you initially set up your router, type them into the login box. If you didn't create a username or password, look for the default ones listed in your wireless router's instructions or manuals.

2. Find the Internet settings, which are usually listed under the WAN (Wide Area Network) heading. The WAN button of a typical configuration screen is shown in Figure 4-1.

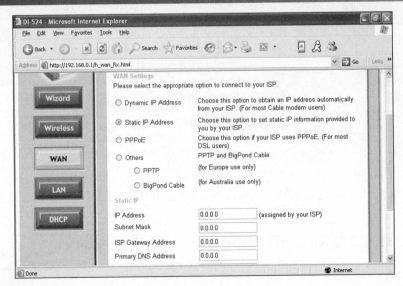

Figure 4-1

3. Choose the Static IP Address option.

4. Enter the IP Address information provided by your Internet provider.

5. Save the changes by clicking either the Apply or Save Settings button located at the bottom of the screen.

Enable DHCP

At first glance, the acronym "DHCP" may seem a bit intimidating, but in actuality it is a simple concept. DHCP—which stands for Dynamic Host Configuration Protocol—is the process in which a wireless router gives a unique IP address to each wireless card or mobile device that wants to access the wireless network. If you are setting up a new wireless router that hasn't been used on a network before, then you won't need to change the DHCP settings. However, if you encounter connection problems while setting up your wireless network or when attempting to access the Internet, it could be the result of your DHCP being disabled. To enable them, do the following:

1. Log in to your wireless router's configuration utility, as follows:

   a. Make sure you are connected to your wireless network. Test this by hovering your mouse pointer over the wireless icon located in the lower-right corner of Windows. A message indicates your connected status.

   b. Open a Web browser like Internet Explorer. If your router is not connected to the Internet, your browser will not find a valid Web page. Instead, it displays an error message that says something like "Page cannot be displayed" or "Page not found."

c. In your Web browser's address window (the place where you usually type the name of a website you want to visit), type your router's IP address. and then press the Enter key. Do not type http:// or www. Instead, only type the number for your IP address.

d. A login box appears. If you created a username and password when initially setting up your router, type them into the login box. If you didn't create a username or password, look for the default ones listed in your wireless router's instructions or manuals.

2. Locate the DHCP settings, which usually appear as a tab you can click.

3. Select the option to enable DHCP.

4. Save the changes by clicking either the Apply or Save Settings button located at the bottom of the screen.

Set the SSID for Your Router

FYI
Choose an SSID that doesn't give clues about where your wireless network is located or who you are.

The SSID (Service Set Identifier) is the name used by wireless routers and cards to identify themselves. By default, most routers have a common SSID that is related to their manufacturer. For example, most Linksys routers use "Linksys" as their SSID. Because these default settings are widely known by criminals, you should change your SSID to help protect your network. You can choose any name you want, so be creative. Here are some examples:

- NewEnglandPatriotsRule
- TheWirelessWonder
- HomeSweetHome

To change the SSID for your router, do the following:

1. Log in to your wireless router's configuration utility, as follows:

a. Make sure you are connected to your wireless network. Test this by hovering your mouse pointer over the wireless icon located in the lower-right corner of Windows. A message indicates your connected status.

b. Open a Web browser like Internet Explorer. If your router is not connected to the Internet, your browser will not find a valid Web page. Instead, it displays an error message that says something like "Page cannot be displayed" or "Page not found."

c. In your Web browser's address window (the place where you usually type the name of a website you want to visit), type your router's IP address, and then press the Enter key. Do not type http:// or www. Instead, only type the number for your IP address.

d. A login box appears. If you created a username and password when initially setting up your router, type them into the login box. If you didn't create a username or password, look for the default ones listed in your wireless router's instructions or manuals.

2. Find the section containing a box that says SSID (or similar). Usually it can be located under the basic wireless settings. Figure 4-2 shows the SSID settings for a Linksys router.

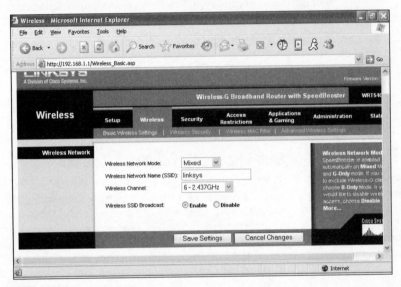

Figure 4-2

3. Type a new SSID.

4. Save the changes by clicking either the Apply or Save Settings button located at the bottom of the screen.

Note

If your computer is currently connected to your router, you will temporarily lose the connection until your wireless card can recognize the new SSID and reconnect to the router. This process is automatic and shouldn't take long.

Disable SSID Broadcasting

By default, most wireless routers broadcast the names of their networks (the SSIDs) over the airwaves so other computer users can see them and possibly log onto the network. This can be a security risk, because it alerts strangers (including criminals) that you have a wireless network just waiting to be hacked into. To help protect yourself, turn off SSID broadcasting. Follow these steps:

1. Log in to your wireless router's configuration utility, as follows:

   a. Make sure you are connected to your wireless network. Test this by hovering your mouse pointer over the wireless icon located in the lower-right corner of Windows. A message indicates your connected status.

   b. Open a Web browser like Internet Explorer. If your router is not connected to the Internet, your browser will not find a valid Web page. Instead, it displays an error message that says something like "Page cannot be displayed" or "Page not found."

   c. In your Web browser's address window (the place where you usually type the name of a website you want to visit), type your router's IP address, and then press the Enter key. Do not type http:// or www. Instead, only type the number for your IP address.

   d. A login box appears. If you created a username and password when initially setting up your router, type them into the login box. If you didn't create a username or password, look for the default ones listed in your wireless router's instructions or manuals.

2. Find the SSID Broadcast option. This may be located in advanced or basic wireless settings or in the advanced performance section (previously shown in Figure 4-2).

3. Check the option to disable SSID broadcasting.

4. Apply the changes by clicking Apply or Save Settings on the bottom of the screen.

It's a Fact

Even if you disable SSID broadcasting, some criminals can use sophisticated software tools to uncover your SSID. For that reason, follow all of the security measures listed in this book.

Note

After disabling SSID broadcasting, you may need to connect manually to your wireless network because it won't be listed as an available network. For more information on making a manual connection, see Chapter 7.

Change the Default Password

To access your router's configuration utility, you must enter a password. By default, most routers are given a common password like "admin." If you don't change your password, you run the risk of having a criminal hijack your network and alter your settings. Here are some things to keep in mind when you create a new password:

• Create a new, unique password. Do not recycle one from an existing account you have with a bank, store, website, and so on.

- Do not use any parts of your name, birthdates, pets' names, or anything else that someone can find out about you by doing a little detective work.

- Use a combination of letters and numbers, and include both uppercase and lowercase letters.

To change your default password, do the following:

1. Log in to your wireless router's configuration utility, as follows:

    a. Make sure you are connected to your wireless network. Test this by hovering your mouse pointer over the wireless icon located in the lower-right corner of Windows. A message indicates your connected status.

    b. Open a Web browser like Internet Explorer. If your router is not connected to the Internet, your browser will not find a valid Web page. Instead, it displays an error message that says something like "Page cannot be displayed" or "Page not found."

    c. In your Web browser's address window (the place where you usually type the name of a website you want to visit), type your router's IP address, and then press the Enter key. Do not type http:// or www. Instead, only type the number for your IP address.

    d. A login box appears. If you created a username and password when initially setting up your router, type them into the login box. If you didn't create a username or password, look for the default ones listed in your wireless router's instructions or manuals.

2. Locate your current password. Usually it can be found in the Administration or Maintenance section (see Figure 4-3).

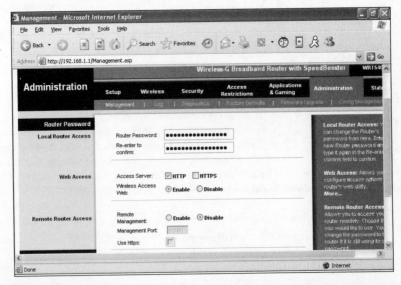

Figure 4-3

3. Type your new password into the box.

4. Save the changes by clicking either the Apply or Save Settings button located at the bottom of the screen.

Configure Advanced Settings

If your wireless network is used at home or in a small office, you probably won't need to change any of the advanced settings in your router (see Figure 4-4 for an idea of what these advanced settings look like). However, if you are a true Geek and want to tweak the performance of your wireless network, follow these guidelines:

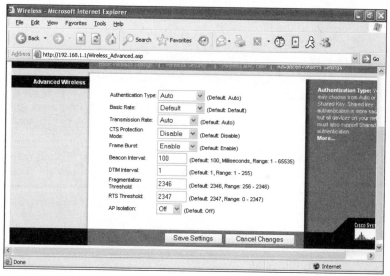

Figure 4-4

- **RTS/CTS (for wireless routers and cards):** If two people are using wireless computers on opposite ends of a large home or office that has a wireless router located in the center of the building, the distance between those computers might cause them to have diffi-culty "hearing" each other. As a result, they may send signals through the air at the same time, causing those signals to collide and produce errors. If both computers try to retransmit the data, the collisions will occur again. To fix this problem, turn on the RTS/CTS feature (which stands for "request to send/clear to send"). Be advised that in order for this to work, RTS/CTS must be enabled on the wireless router and on the wireless cards in each computer. For more infor-mation on RTS/CTS, see Chapter 10.

- **Fragmentation (for wireless routers and cards):** If your wireless network is having difficulty sending or receiving data, it might be due to interference from other radio signals in the air. To fix this problem, turn on the Fragmentation feature in your router and in your wireless cards. For more information on Fragmentation, see Chapter 10.

- **Beacon Interval (for wireless routers only):** If you are using a laptop computer or mobile device and want to extend its battery's life, try increasing the Beacon Interval in your wireless router. But be warned: this could cause your wireless network to slow down.

- **DTIM Period (for wireless routers only):** Another way to conserve power in your mobile devices is to turn on a feature called DTIM Period. But be warned: it could cause your wireless network to slow down.

- **Data Rates (for wireless routers and cards):** If you always use your mobile computer near your wireless router, consider changing your router's data rate to its highest setting (usually 54 Mbps). This will give you a steady, fast wireless signal. But be warned: if you increase the data rate and try to use your wireless computer too far away from your router, you may have problems sending and receiving signals. On the flip side, if you want to use your wireless computer as far away from your router as possible, consider changing the data rate to a very slow setting (like 1 Mbps).

- **Protection Mechanisms (for wireless routers only):** If you have a large wireless network with many users, consider turning on the Protection Mechanisms feature. This will prevent computers using the 802.11b wireless cards from interfering with computers using the 802.11g cards.

- **802.11g Only Mode (for wireless routers only):** If all of the wireless cards in your computers are 802.11g, then you can set your wireless router to "802.11g only" mode. This will prevent anyone with the slower 802.11b devices from connecting to your wireless network and slowing it down.

- **Authentication (for wireless routers and cards):** Turning on this feature helps prevent criminals from sneaking into or hijacking your wireless network. There are several types of authentication: open systems, shared key, WEP, and WPA. Don't use shared key, because it is easy for criminals to crack. Your best bet is to use WPA. If that option is not available, then use WEP. If you are setting up a wireless network known as a "hotspot" that will be used by the general public (similar to the free wireless networks found in many coffee shops or bookstores), then use the open systems authentication. That way anyone with a wireless computer or mobile device can properly access your network.

5

Configuring Your Wireless Card

Usually you don't have to configure a wireless card in order to access a wireless network. However, you should activate the card's security features to protect your private data. Here is a general overview of those features:

- **Encryption method:** To prevent a criminal from snatching your wireless signals out of the air and reading your e-mail or viewing your private information, you must turn on an encryption mode in your wireless card. For wireless networks in home and small office networks, use WPA-PSK. If that's not available, use WEP.

- **Network Name (SSID):** In most cases, Windows XP will display a list of network names (their SSIDs) that you can access. If not, you need to manually enter the SSID.

- **Wireless Channel:** If you are using an ad hoc network, you need to assign a specific channel to your wireless card. For more information on using ad hoc networks, see Chapter 2.

Enable Your Wireless Card

Before you can connect to a wireless network, you must make sure your wireless card has been properly enabled. Once the card is operating properly, it continues to do so unless you disable it or turn off your computer. Follow these steps to enable your card:

1. Click the Start button in the lower-left corner of Windows.

2. Click Connect To (see Figure 5-1). If this option isn't available, your Start menu is in classic mode. In that case, click Settings, and then select Network Connections.

Do It Yourself

Enable your wireless card

Disable your wireless card

Change your wireless card's settings

Access the configuration utility

Enable or disable DHCP

Find the IP address for your wireless card

Set a static IP address for your wireless card

Set computer network identification

Turn on power-save mode

Figure 5-1

3. Click the Wireless Network Connection you want to enable.

Disable Your Wireless Card

If you are not currently connected to a wireless network and don't have plans to do get connected any time soon, you can increase your computer's security and extend the life of its battery (for laptops or mobile devices) by disabling your wireless card. Here's how:

1. Double-click the wireless icon in the lower-right corner of Windows. The Wireless Network Connection Status Dialog box opens (see Figure 5-2).

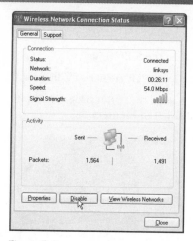

Figure 5-2

2. Click the Disable button. Your wireless card shuts down until you re-enable it.

Here is an alternate method of disabling a wireless card:

1. Right-click the wireless icon in the lower-right corner of Windows.

2. Select Disable.

Change Your Wireless Card's Settings

Whether you are using a wireless network at your home, office, or in a public setting like a bookstore or hotel, you need to know how to change or tweak your wireless card's settings in order to make your wireless experience satisfying and safe.

Access the Configuration Utility

To change the settings and features in your wireless card, you need to use a program known as a configuration utility. There are two types: those provided by a manufacturer and the one built into Windows XP.

Manufacturer's Configuration Utility

Whether your wireless card came with your computer or you installed it yourself, it probably has a configuration utility designed specially for it. Use this utility whenever you want to change or tweak your card's features. Here's how to access it:

1. Double-click the wireless icon located in the lower-right corner of Windows. The configuration utility opens.

2. If you can't find the wireless icon, do the following:

   a. Click the Start button in the lower-left corner of Windows.

   b. Click All Programs, and then scan the list of programs to find your card's configuration utility (see Figure 5-3).

Figure 5-3

   c. Double-click the program's name to start it (see Figure 5-4).

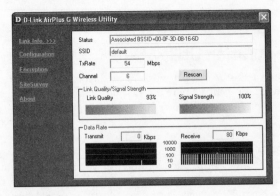

Figure 5-4

Windows XP's Configuration Utility

If your wireless card did not come with a configuration utility, you can use the one supplied by Windows XP. Even if you do have a manufacturer's utility, there may come a time when you need to access the one built into Windows. Here's how:

1. Double-click the wireless icon located in the lower-right corner of Windows.

2. Click Properties.

3. Click the Wireless Networks tab. If this tab isn't available, your wireless card doesn't support this feature.

4. Based on your needs, either check or uncheck the Use Windows to Configure My Wireless Network Settings option (see Figure 5-5).

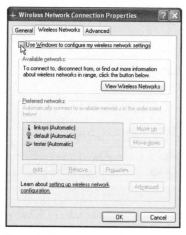

Figure 5-5

Enable or Disable DHCP

DHCP is the feature that gives an IP address to a wireless device (for an in-depth discussion of DHCP and IP addresses, refer to Chapter 4). By default, most wireless cards have DHCP automatically enabled, so you don't have to change anything. However, if you ever encounter a situation in which you need to turn DHCP off or back on again, here's how to do it:

1. Double-click the wireless icon located in the lower-right corner of Windows.

2. Click Properties.

3. In the box, click Internet Protocol (TCP/IP). Be sure not to accidentally check or uncheck this option.

4. Click Properties.

5. Select the Obtain an IP Address Automatically option. If you ever want to disable DHCP on the card, unselect this option. Figure 5-6 shows the wireless card's DHCP settings.

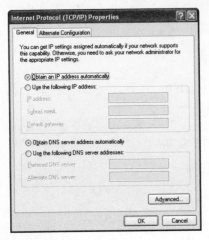

Figure 5-6

Find the IP Address for Your Wireless Card

When using filtering methods or other advanced features of your wireless router, you may need to find out what IP address your wireless card has been given. Here's how:

1. Double-click the wireless icon in the lower-right corner of Windows.

2. Click the Support tab.

3. View the IP Address value (see Figure 5-7). If the value is blank, your card has not yet been given an IP address. Wait a few minutes and try again, or click the Repair button.

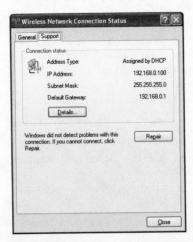

Figure 5-7

Set a Static IP Address for Your Wireless Card

If you must use a static, unchanging IP address in your wireless router (which is required by some Internet providers), then you must also give a static IP address to the wireless card in each of your computers. For more information on static IP addresses, refer to Chapter 4.

Follow these steps to set a static IP address:

1. Double-click the wireless icon in the lower-right corner of Windows.

2. Click Properties.

3. In the box, click Internet Protocol (TCP/IP). Be sure not to accidentally check or uncheck this option.

4. Click Properties.

5. Select the Use the Following IP Address option.

6. In the IP Address box, type the IP address for your wireless card. For example, if your wireless router has a static, unchanging IP address of 192.168.1.1, you would give your first wireless computer an IP address of 198.168.1.2, then give the second computer an address of 192.168.1.3, and so on.

7. In the Subnet Mask box, type the subnet mask for your network (for homes and small offices, 255.255.255.0 works fine).

Note

A subnet mask indicates which part of the IP address the router uses when sending data to a wireless device. A subnet mask of 255.255.255.0 indicates the router will use only the last digit of the IP address for routing purposes.

8. In the Default Gateway box, type the IP address of your wireless router. Figure 5-8 shows the Internet Properties box set to enable static IP addresses. For more information on finding the static IP address of your router, refer to Chapter 4.

9. Click OK to save your changes.

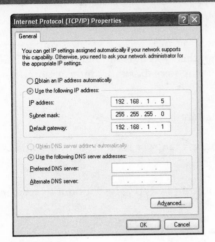

Figure 5-8

Set Computer Network Identification

A feature known as "computer network identification" helps the computers on a network to identify each other when using a common printer, sharing files, or performing other common network activities. It is advisable to turn on this feature in each computer using your wireless network. Here's how:

1. Right-click the My Computer icon on your desktop. If this icon is not available, click the Start button in the lower-left corner of Windows and right-click My Computer. If you can't find the My Computer icon anywhere, do the following:

   a. Right-click in the empty space on your desktop.

   b. Select Properties.

   c. A window opens. Click the Desktop tab.

   d. Near the bottom of the window, click the Customize Desktop button.

   e. Another window opens. On the General tab, beneath Desktop Icons, place a checkmark in the My Computer box.

   f. Click the OK button.

   g. You are returned to the previous screen. Click the Apply button.

   h. Click the OK button.

   i. The My Computer icon appears on your desktop. Right-click it.

2. Select Properties.

3. Click the Computer Name tab. The System Properties box opens and displays the computer's description, name, and the workgroup it belongs to (see Figure 5-9).

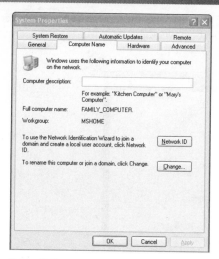

Figure 5-9

4. To set the computer's name and workgroup, click Change.

Note
Workgroups are used primarily in homes and small offices (most larger companies use "domains"). All of your computers should belong to the same workgroup.

5. Enter the new information in the boxes provided, and then click OK.

6. If you want to change your computer's description, type a new one in the corresponding box, and then click OK to save your changes.

Turn On Power-Save Mode

Most wireless devices consume a great amount of power. By using the power-save mode on your laptop's wireless card, you can make your batteries last longer. When power-saving is enabled, your wireless card "sleeps" (temporarily shuts off) when not in use. This sleeping time, even in short intervals, can help to conserve the batteries. There are two ways to turn on the power-save mode: through the manufacturer's configuration utility or through Windows XP.

Use the Manufacturer's Configuration Utility
Follow these steps to turn on power-save with the configuration utility:

1. Double-click the wireless icon in the lower-right corner of Windows.

If this icon is not available, click the Start button in the lower-left corner of Windows and click All Programs to view the list of programs on your computer. Locate and click the name of your wireless card's configuration utility.

2. After the configuration utility opens, find the Power Mode or Power Save Mode setting. Figure 5-10 shows the Power Mode settings for the D-Link AirPlus G wireless card.

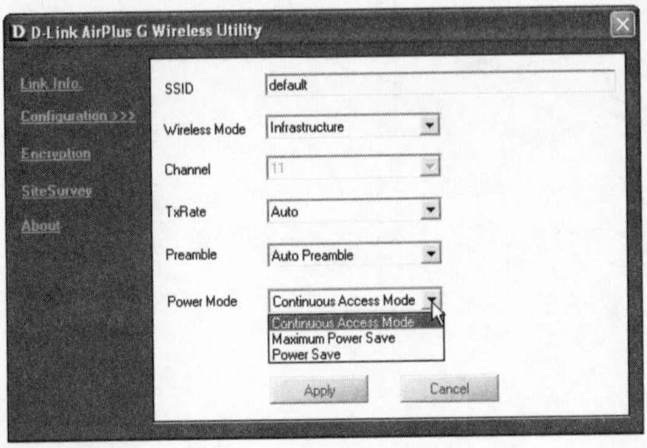

Figure 5-10

3. Click the appropriate box or button to enable the power-save mode.

Use Windows XP Home Edition and XP Professional Edition

Follow these steps to turn on power-save through Windows XP:

1. Double-click the wireless icon in the lower-right corner of Windows, as follows:

If this icon is not available, click the Start button in the lower-left corner of Windows and click All Programs to view the list of programs on your computer. Locate and click the name of your wireless card's configuration utility.

2. Click Properties.

3. Click Configure.

4. Click the Advanced tab (see Figure 5-11). If the Advanced tab isn't available, then unfortunately you can't use Windows to change your wireless card's settings. In that case, you need to use the manufacturer's configuration utility (described earlier).

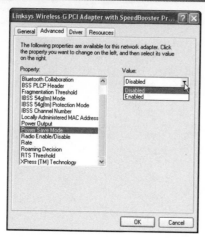

Figure 5-11

5. In the list of properties located on the left, click Card Power Management or Power Save Mode.

6. In the Value drop-down list, select your desired setting. Some wireless cards may have settings named Continuous Access Mode, Maximum Power Save, or Power Save. However, other wireless cards only let you select Enabled and Disabled.

7. Click OK. Your wireless card will now use the power-save mode you selected.

6

Security Settings

I f your computer has ever been infected with a virus, worm, or spyware, then you know firsthand the kind of digital dangers lurking on the Internet. But these aren't the only ones you must worry about. If you use a wireless computer or a wireless network, here are some specific threats to your safety:

- **Wi-Fi hackers:** These criminals drive through neighborhoods and business districts to see if they can access an unprotected or unencrypted wireless computer network. If successful, they can hijack that wireless Internet connection to send spam or download illegal materials (like child pornography). Also, the intruders are only one step away from being able to hack into the victim's computer, view private files — in particular credit card numbers and social security numbers — and use that information to commit identity theft.

- **Evil twin hotspots:** In computer terms, an *evil twin* is a free, wireless hotspot created by a criminal to mimic the public Internet access available at places like coffeeshops and bookstores. Usually the evil twin hotspot looks legitimate, so customers connect to it and send e-mail, surf the Web, and do online banking without any idea that everything they are doing or typing is being recorded by the criminal (including passwords and bank account numbers). To avoid becoming a victim of an evil twin hotspot, see the "Disable Windows' Search for Any Available Network" 5-minute fix later in this chapter.

- **DoS attacks:** In a denial of service attack (DoS), a criminal attempts to knock a website off the Internet or crash a computer network. This can be done by using special hacking software or by using an army of computers that have fallen under the criminal's control after being infected with a digital virus or worm (often these hijacked computers are called "zombies" or "drones").

Do It Yourself

Protect your wireless computers and network

Protect credit cards over a wireless connection

Stay safe when using public networks

Disable Windows' search for any available network

Wireless encryption

Turn on WPA-PSK encryption

Turn on WEP encryption

So with all of this criminal activity taking place, is it even safe to use wireless computers? Yes—but only if you follow all of the security advice in this chapter (and in the rest of this book).

Protect Your Wireless Computers and Network

No matter whether you are worried about hackers, evil twins, or any other wireless crime, there are several simple things you can do to protect yourself, including the following:

- **Use WEP or WPA-PSK encryption:** This critical security feature requires computer users to enter a password before they are allowed to access your wireless network. In addition, it protects all of the information you send over the airwaves by encrypting it (which makes it virtually unreadable to hackers). For more information, see the "Turn on WPA-PSK Encryption" and "Turn on WEP Encryption" 5-minute fixes later in this chapter.

- **Reduce the range of your network:** To reduce the possibility of a nosy neighbor or hacker accessing your wireless network without permission, you should place your router in the center of your home or office and reduce the transmit power. By default, most wireless routers have their transmit power set at the highest level—100 percent—but allow you to change it to a lower value (see Figure 6-1). Try decreasing the transmit power one setting at a time, then check to see if all computers using your wireless network can still send and receive data at an acceptable speed. To learn how to do this, refer to Chapter 9.

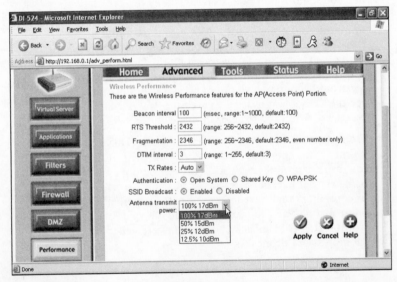

Figure 6-1

- **Change your router's default password:** To access your router's configuration utility, you must enter a password. By default, most routers are given a common password like "admin." If you don't change your password, you run the risk of having a criminal hijack your network and alter your settings. For instructions on changing your password, refer to Chapter 4.

- **Disable SSID broadcasting:** By default, most wireless routers broadcast the names of their networks (the Service Set Identifiers or SSIDs) over the airwaves so other computer users can see them and possibly log onto the network. This can be a security risk, because it alerts strangers (including criminals) that you have a wireless network just waiting to be hacked into. To help protect yourself, turn off SSID broadcasting. To learn how to do this, refer to Chapter 4.

- **Change the default SSID of your wireless router:** The SSID (Service Set Identifier) is the name used by wireless routers to identify themselves. By default, most routers have a common SSID that is related to their manufacturer. For example, most Linksys routers use "Linksys" as their SSID. Because these default settings are widely known by criminals, you should change your SSID to help protect your network. To learn how to do this, refer to Chapter 4.

Protect Credit Cards over a Wireless Connection

One of the biggest concerns people have about wireless devices is whether it is safe to use a credit card when making an Internet purchase. Fortunately you have little to worry about as long as you follow these security tips:

- **Use encryption:** To protect your credit card number from being snatched out of the air, turn on the encryption features of both your wireless router and your wireless cards. The best encryption method to use is WPA-PSK. If that is not available, then use WEP.

- **Look for the padlock:** Before entering your credit card number on a website, make sure that site is using SSL encryption. This is indicated by an icon resembling a padlock located in the status bar at the bottom of your Internet browser (see Figure 6-2).

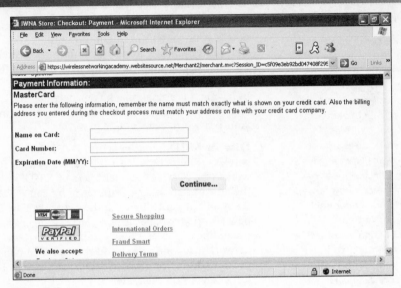

Figure 6-2

- **Don't use debit or check cards:** When shopping online, never use a debit card, check card, or any other card that links directly to your checking account. This protects your account from being wiped clean by an Internet criminal who steals your identity. Although most debit or check cards claim to provide 100-percent protection in the event of identity fraud, it could take several hours or days for the money to be replaced in your checking account (especially if you have to fight with the bank to prove you didn't make the fraudulent purchases). During that time, it is possible that checks you recently wrote could bounce, and current credit transactions could be declined. For maximum protection, use a standard credit card that is not connected to your bank accounts.

- **Use one card**— To make it easier to keep track of your Internet purchases and to spot fraud, use only one credit card for all of your e-commerce.

- **Use virtual accounts**— Some credit-card companies offer "virtual account numbers" that link directly to your credit card and expire as soon as they are used. By using a different virtual account each time you make an Internet purchase, you will protect your real credit card number from landing in the hands of high-tech crooks.

Stay Safe When Using Public Networks

Public wireless networks (also known as hotspots) don't use encryption, which means your data is not protected while being transmitted through the air. It is possible for a criminal using high-tech equipment to intercept this data and view your user names, passwords, credit card numbers, financial documents, or other private information. To stay safe while using a hotspot, follow these tips:

- **Disable your wireless card when not using it:** There is no better way to secure your computer than shutting off its wireless card when you don't need to use it. For details on how to do this, refer to Chapter 5.

- **Use a firewall:** A firewall is a program that acts like a shield to protect your computer from the watchful eyes of Internet criminals. It also can filter the data that enters your computer, control Internet cookies, and warn you when sinister spyware programs try to transmit data about you over the Web.

- **Use SSL encryption:** If you see an icon resembling a padlock in the lower-right corner of your Internet browser, then you can safely enter credit-card numbers or passwords into the current Web page.

- **Don't use shared folders:** Before accessing a public wireless network, remove important files from your Shared Documents folder. Also, you should disable file sharing in your folders by doing the following:

    1. Right-click the folder you no longer want to share.

    2. Choose Properties.

    3. Select Sharing.

    4. Remove the checkmark from the Share This Folder on the Network box (see Figure 6-3).

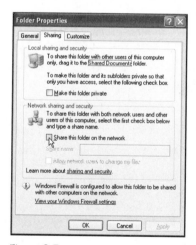

Figure 6-3

    5. Click the Apply button.

    6. Click the OK button.

- **Watch for wandering eyes:** An easy way for a thief to steal your passwords or other sensitive information requires no technology: all he or she has to do is look over your shoulder while you type. Keep an eye out for these snoops, and consider turning down the brightness or intensity of your computer screen.

Disable Windows' Search for Any Available Network

Many wireless computers have a setting in Windows that enables them to connect to the nearest available wireless network that is providing a strong signal. If you are using this feature and your computer has difficulty receiving a steady signal from your own network, it may switch to someone else's network nearby. Also, this is the same feature that allows unsuspecting computer users to become victims of evil twin hotspots (as described in the beginning of this chapter). To enhance your security and give yourself peace of mind, disable this feature, as follows:

1. Double-click the wireless icon located in the lower-right corner of Windows.

2. Click Properties.

3. Click the Wireless Networks tab.

4. Click the Advanced button. If this button is grayed out or is not available, then you aren't using Windows to configure your wireless settings. In that case, check the manufacturer's configuration utility that came with your wireless card to see if it has a similar setting that must be turned off. For more information on using the manufacturer's configuration utility, refer to Chapter 5.

5. Uncheck the Automatically Connect to Non-preferred Networks option (see Figure 6-4).

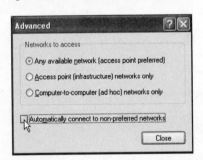

Figure 6-4

6. Click Close.

Wireless Encryption

To keep unwanted intruders from viewing the information you transmit over the airwaves, turn on an encryption mode like WPA or WEP. Most older wireless routers only use WEP, but almost all new ones have WPA. If you have a choice between them, use WPA because it is stronger.

Turn on WPA-PSK Encryption

Follow these steps to turn on WPA-PSK encryption.

Note

You may need to download an update to make Windows support WPA-PSK. To learn how to do this, see Chapter 9.

1. Log in to your wireless router's configuration utility, as follows:

a. Make sure you are connected to your wireless network. Test this by hovering your mouse pointer over the wireless icon located in the lower-right corner of Windows. A message should indicate your connected status.

b. Open a Web browser like Internet Explorer. If your router is not connected to the Internet, your browser will not find a valid Web page. Instead, it displays an error message that says something like "Page cannot be displayed" or "Page not found."

c. In your Web browser's address window (the place where you usually type the name of a website you want to visit), type your router's IP address, and then press the Enter key. Do not type http:// or www. Instead, only type the number for your IP address.

d. A login box appears. If you created a username and password when initially setting up your router, type them into the login box. If you didn't create a username or password, look for the default ones listed in your wireless router's instructions or manual.

2. Locate the router's encryption settings, which are usually in the Wireless Security (or similarly named) section (see Figure 6-5).

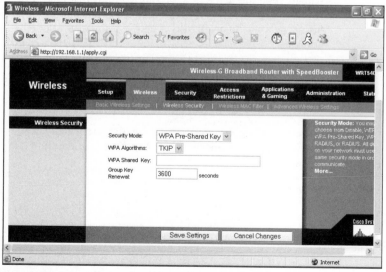

Figure 6-5

3. Select WPA-PSK. However, if there is an option to select an encryption mode called AES (Advanced Encryption Standard), use it because it offers stronger encryption than WPA.

4. Create an encryption key (a type of password) that contains between 8 and 63 characters. These characters should be a combination of letters, numbers, and symbols (for example, N1o2S0i8D0a5M). You will be required to type it into each wireless computer that wants to access your network, so don't lose or forget the key. Either memorize it or write it down and store it in a secure location like a fire safe.

5. Save the changes by clicking either the Apply or Save Settings button located at the bottom of the screen.

Turn on WEP Encryption

If your wireless router doesn't offer WPA encryption, then WEP can be used as an acceptable alternative. Follow these steps to turn on WEP encryption:

1. Log in to your wireless router's configuration utility, as follows:

    a. Make sure you are connected to your wireless network. Test this by hovering your mouse pointer over the wireless icon located in the lower-right corner of Windows. A message should indicate your connected status.

    b. Open a Web browser like Internet Explorer. If your router is not connected to the Internet, your browser will not find a valid Web page. Instead, it displays an error message that says something like "Page cannot be displayed" or "Page not found."

    c. In your Web browser's address window (the place where you usually type the name of a website you want to visit), type your router's IP address, and then press the Enter key. Do not type http:// or www. Instead, only type the number for your IP address.

    d. A login box appears. If you created a username and password when initially setting up your router, type them into the login box. If you didn't create a username or password, look for the default ones listed in your wireless router's instructions or manual.

2. Locate the router's encryption settings, which are usually in the Wireless Security (or similarly named) section (see Figure 6-6).

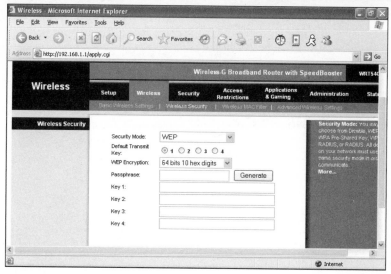

Figure 6-6

3. Select WEP, and then choose your desired type of WEP encryption: 64-bit or 128-bit. If 128-bit is available, use it because it offers more security.

4. If there is an option for a key type, select HEX or ASCII (either works fine).

5. If there is an option to create a passphrase, then type a phrase and click the Generate option. Four WEP encryption keys are automatically generated for you.

FYI

To ensure the continued security of your wireless network, change your WEP key frequently.

6. Select the WEP key you want to use. You will be required to type it into each wireless computer that wants to access your network, so don't lose or forget the key. Either memorize it or write it down and store it in a secure location like a fire safe.

7. Save the changes by clicking either the Apply or Save Settings button located at the bottom of the screen.

PART IV

USING AND MAINTAINING YOUR NETWORK

After installing a wireless network, you might need some help using it or access-ing some of its important features. Not only does this section of the book show you the best way to use a wireless network, but it also discusses some mainte-nance tasks that should be performed periodically to keep your wireless network running smoothly.

7

Using Your Network

You've done all the hard work. Your wireless cards are installed. Your router is configured. Now it's time to start using your network.

Test Your Network

In most cases, a single wireless router can provide a strong signal to an entire home or small office (assuming you placed the router near the center of the building). If you have a wireless laptop, you can do some testing to find out whether your router is sitting in the ideal spot. Here's how:

1. With your laptop in hand, walk around various areas of your home or office to see if you can maintain a connection. To do so, observe the signal strength and connection status of your wireless card, hovering your mouse over the wireless icon in the lower-right corner of Windows. A pop-up message appears with connection status information.

2. If you have difficulty connecting to your network in a particular area, try moving the router closer to that area. Next, rerun the connection tests to see if the signal strength has improved.

3. If you can't find an optimum location for your router that provides full wireless coverage throughout your entire home or office, consider installing a second router.

Common Networking Tasks

Once your wireless network is in place and operating properly, you may need help getting the most out of it. Here is an overview of some common networking tasks you might need to do on a regular basis.

Do It Yourself

Test your network

Common networking tasks

Automatically connect to your wireless network

Manually connect to a wireless network

Connect to a public wireless network

Renew your IP address

Share files in Windows XP

Access shared folders

Share files in older versions of Windows

Filter content on your network

Automatically Connect to Your Wireless Network

The one task you will probably do more than any other is connecting your computer to a wireless network by using the automatic connection features of Windows XP or your wireless card's configuration utility.

Use Windows XP Home Edition and XP Professional Edition

If you are using Windows XP, the easiest way to connect to a wireless network is to view a list of available networks and select the one you want to access. Follow these steps:

1. Right-click the wireless icon located in the lower-right corner of Windows.

2. Click View Available Wireless Networks. The Wireless Network Connection box opens and displays a list of the available networks (see Figure 7-1).

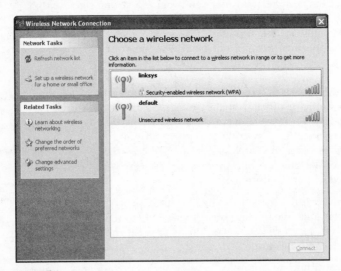

Figure 7-1

3. If you don't see any available wireless networks, make sure your wireless router is working properly. If you still can't see any available wireless networks, it may be because Windows XP hasn't been given permission to configure and oversee your wireless settings. Here's how to do that:

    a. Double-click the wireless icon in the lower-right corner of Windows.

    b. Click Properties.

    c. Click the Wireless Networks tab.

    d. Check the Use Windows to Configure My Wireless Network Settings option (see Figure 7-2).

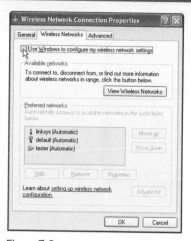

Figure 7-2

4. Click the wireless network you want to access. If you don't have
Windows XP's Service Pack 2 installed and you are trying to connect
to an unsecured wireless network, you may have to check the option
that says, "Allow me to connect to the selected wireless network, even
though it is not secure."

5. Click Connect. If you have Windows XP's Service Pack 2 installed and
the wireless network you are trying to connect to is secured, you are
asked to type an encryption key or passphrase (see Figure 7-3).

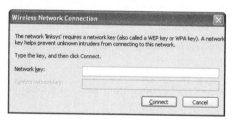

Figure 7-3

Use the Manufacturer's Configuration Utility

If you are using the manufacturer's configuration utility that came with your
wireless card, the easiest way to connect to a wireless network is to view a list of
available networks and select the one you want to access. Here's how:

1. Double-click the wireless icon located in the lower-right corner of
Windows.

2. Click the tab that shows the available wireless networks (see Figure 7-4).

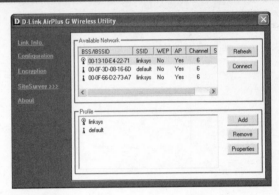

Figure 7-4

3. Select the wireless network you want to access.

4. Click Connect.

5. If the wireless network is secured, you may be required to type an encryption key or passphrase.

Manually Connect to a Wireless Network

If you ever need to access a wireless network that uses security features to hide itself from the public's prying eyes, then follow the steps in this section to make a manual connection.

Note

To make a manual connection, you must know the name of the network (its SSID). For more information on SSIDs, refer to Chapter 4.

To Use Windows XP Home Edition or XP Professional Edition to manually connect:

1. Right-click the wireless icon in the lower-right corner of Windows.

2. Click View Available Wireless Networks.

3. If you have Windows XP's Service Pack 2 installed, click the Change Advanced Settings option. If you don't have Windows XP's Service Pack 2 installed, click Advanced.

4. Click the Wireless Networks tab (see Figure 7-5). If this tab isn't available, you must use the manufacturer's configuration utility that came with your wireless card (as described next).

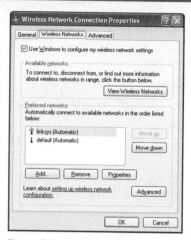

Figure 7-5

5. Click Add.

6. Type the name (the SSID) of the wireless network you want to connect to (see Figure 7-6).

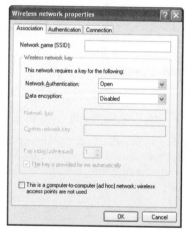

Figure 7-6

7. If the wireless network has encryption enabled, select the encryption type and enter the encryption key or passphrase.

8. Click OK.

To use the manufacturer's configuration utility to manually connect:

1. Double-click the wireless icon in the lower-right corner of Windows. If this icon is not available, do the following:

   a. Click the Start button in the lower-left corner of Windows.

   b. Click All Programs to display a list of the programs on your computer.

   c. Click the name of the manufacturer's configuration utility for your wireless card.

2. After the configuration utility opens, find the section containing your wireless settings. Often it is located on a Configuration tab (see Figure 7-7).

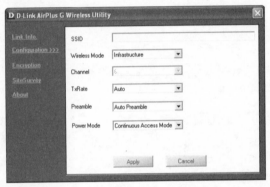

Figure 7-7

3. Type the name (the SSID) of the wireless network you want to access.

4. If the wireless network has encryption enabled, select the encryption type and enter the encryption key or passphrase. Some manufacturers have a separate tab for entering the encryption information.

5. Click Apply to save the changes.

Connect to a Public Wireless Network

Using a public wireless network—often called a "hotspot" or a "Wi-Fi zone"—is a convenient way to check e-mail or browse the Internet when you are away from home or the office. Due to the rising popularity of these hotspots, increasing numbers of stores like coffeehouses or book-sellers are making them available free of charge. If you are interested in accessing a hotspot, do the following:

1. Find a public hotspot by looking for Wi-Fi signs at stores, asking around, or visiting websites like http://www.wi-fizone.org to pinpoint hotspots in your area.

FYI

Another way to search for public hotspots is to use a "Wi-Fi finder." This handy device— which is small enough to fit on your keychain—will beep or flash some lights when you are in range of a wireless network. You can find Wi-Fi finders at most computer or electronic stores for under $50.

2. Before you can sign into a public wireless network you must first connect to one of the wireless routers in use by the network. To do this, you need to know the SSID of the network. This may be obvious, but in some cases, you may need to ask. Once you know the SSID, proceed with connecting to that network.

FYI

When paying for wireless access at a hotspot, make sure the website that accepts your payment uses SSL encryption. This protects your credit-card number from being snatched out of the air by high-tech criminals. To determine if the proper encryption is being used, look for an icon that looks like a padlock located in the lower-right corner of your Web browser. If it's there, you're safe.

3. Open your Web browser. It probably won't show your usual home page because most public hotspots automatically redirect you to their own Web page. With free hotspots, this Web page may only ask you to accept some general terms and conditions. If you find those terms to be agreeable, go ahead and accept them. In some places, like hotels, you might also need to enter your room number or a password given to you by the check-in desk. With hotspots that require you to pay a small fee, the Web page will offer you several service plans. These may be per-minute, per-hour, daily, or monthly plans. Choose the plan that makes most sense to you, and then pay by entering your credit card information.

4. After you pay for the service, you'll probably need to log in. Type the user name and password you were given when subscribing to the service.

5. After logging in, you can use e-mail or surf the Internet like normal.

Renew Your IP Address

If you move your wireless computer or mobile device from one wireless network to another, you may find that you cannot access the new network or use the Internet. In that case, you can usually solve the problem by renewing your IP address.

To renew your IP address with Windows XP Home Edition and XP Professional Edition:

1. Double-click the wireless icon in the lower-right corner of Windows.

2. Click the Support tab.

3. Click Repair (see Figure 7-8). This renews your IP address.

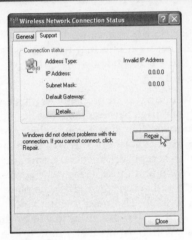

Figure 7-8

Here's an alternate method to renew your IP address with Windows XP Home Edition and XP Professional Edition:

1. Click the Start button in the lower-left corner of Windows.

2. Select All Programs.

3. Click Accessories.

4. Click the Command Prompt icon.

5. A command prompt window opens. Type **ipconfig /renew** in the window (see Figure 7-9), and then press the Enter key. This renews your IP address and displays it on your screen.

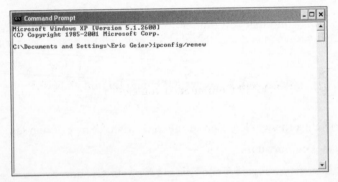

Figure 7-9

Share Files in Windows XP

Sharing files on your network can be a great convenience. By default, Windows XP is configured to share files through your Shared Documents folder. In addition, you can share other folders of your choosing.

Share Individual Files

To share only a few carefully-selected files, do the following:

1. Double-click the My Computer icon on your desktop. If this icon is not available, click the Start button in the lower-left corner of Windows and click My Computer. If you can't find the My Computer icon anywhere, do the following:

    a. Right-click in the empty space on your desktop.

    b. Select Properties.

    c. A window opens. Click the Desktop tab.

    d. Near the bottom of the window, click the Customize Desktop button.

    e. Another window opens. On the General tab, beneath Desktop Icons, place a checkmark in the My Computer box.

    f. Click the OK button.

    g. You are returned to the previous screen. Click the Apply button.

    h. Click the OK button.

    i. The My Computer icon appears on your desktop. Double-click it.

2. Double-click the Shared Documents folder. This is the place where you will store the files you are willing to share with other computers connected to your network. You may drag (or move), copy, or create files in this folder to be shared on the network.

Share an Entire Folder

To share the entire contents of a particular folder on your computer, do the following:

1. Double-click the My Computer icon on your desktop. If this icon is not available, click the Start button in the lower-left corner of Windows and click My Computer. If you can't find the My Computer icon anywhere, do the following:

    a. Right-click in the empty space on your desktop.

    b. Select Properties.

    c. A window opens. Click the Desktop tab.

    d. Near the bottom of the window, click the Customize Desktop button.

    e. Another window opens. On the General tab, beneath Desktop Icons, place a checkmark in the My Computer box.

    f. Click the OK button.

    g. You are returned to the previous screen. Click the Apply button.

    h. Click the OK button.

    i. The My Computer icon appears on your desktop. Double-click it.

2. Double-click the icon for your C: drive.

3. Browse through the contents of your C: drive until you find a folder you want to share (see Figure 7-10).

Figure 7-10

4. Right-click the folder and select Sharing and Security.

5. Check the Share This Folder on the Network option (see Figure 7-11).

6. Type a "share name" in the provided text box. Any name will do.

7. If you want to permit other people to make changes to your shared files, check the Allow Network Users to Change My Files option.

FYI

If you are sharing folders other than the one labeled Shared Documents, you should turn on the security features of your wireless router and wireless cards. This prevents strangers (including criminals) from connecting to your network and viewing or altering your shared files. For more information on enabling these security features, refer to Chapters 4, 5, and 6.

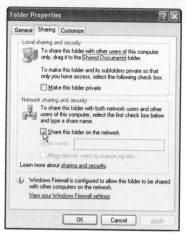

Figure 7-11

Access Shared Folders

Regardless of whether or not you want to share your files or folders, you can view the shared folders that other people on your wireless network have made available. Here's how.

Note

To access someone's shared folders, that person's computer must be running.

1. Double-click the My Network Places icon on your desktop. If this icon is not available, do the following:

   a. Click the Start button in the lower-left corner of Windows.

   b. Click All Programs.

   c. Click Accessories.

   d. Click Windows Explorer.

   e. In the left window pane, click My Network Places (see Figure 7-12).

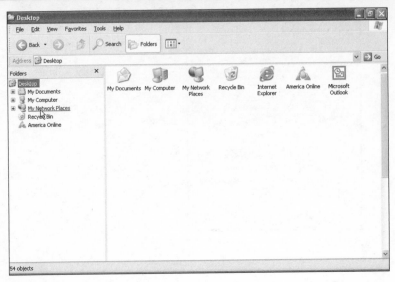

Figure 7-12

2. In the Local Network section, double-click the shared folder you are looking for.

Share Files in Older Versions of Windows

The sharing of files and folders in older versions of Windows (like 98 or ME) is handled differently than the method used by the modern version of Windows (called XP).

To enable file sharing in Windows 98 and Windows ME:

1. Double-click the My Computer icon on your desktop.

2. Double-click the Control Panel icon.

3. Double-click the Network icon.

4. Click the File and Print Sharing button. The File and Print Sharing window appears (see Figure 7-13).

Figure 7-13

FYI

Occasionally you may encounter minor problems when sharing files across a wireless network. For example, when attempting to open a very large file located in a shared folder on someone's computer, you might find that the file seems to take forever to open. A fast, easy way to avoid this difficulty is to copy the file to your own computer and open it there.

5. Check the option that says, "I want to be able to give others access to my files."

6. Click the OK button.

To access shared files on a network in Windows 98 and Windows ME:

1. Double-click the Network Neighborhood icon on your desktop to open the Network Neighborhood window (see Figure 7-14).

Figure 7-14

2. Find and double-click the icon of a computer name. You should be able to see the computer's shared folders and printers.

Filter Content on Your Network

Most wireless routers have filtering features that allow you to block particular computers from accessing your network at specified times. In addition, you can block specific websites and limit access to services like POP3 e-mail, AOL Instant Messenger, Yahoo Messenger, and FTP file transfers. For example, you may want to prevent your child from accessing the Internet between the hours of 10 PM and 8 AM each day. Here is a general overview of the steps involved:

1. Log in to your wireless router's configuration utility.

   a. Make sure you are connected to your wireless network. Test this by hovering your mouse pointer over the wireless icon located in the lower-right corner of Windows. A message should indicate your connected status.

b. Open a Web browser like Internet Explorer. If your router is not connected to the Internet, your browser will not find a valid Web page. Instead, it displays an error message that says something like "Page cannot be displayed" or "Page not found."

c. In your Web browser's address window (the place where you usually type the name of a website you want to visit), type your router's IP address, and then press the Enter key. Do not type http:// or www. Instead, only type the number for your IP address.

d. A login box appears. If you have created a username and password when initially setting up your router, type them into the login box. If you didn't create a username or password, look for the default ones listed in your wireless router's instructions or manual.

FYI

To block specific Internet services, you may need to know their port numbers. Here are some common services and their corresponding ports: HTTP (Internet) uses port 80, FTP (File Transfer) uses ports 20-21, and POP3 e-mail uses port 110.

2. Find the filtering settings. Often they are located in a Filters or Access Restriction section (see Figure 7-15).

3. Set your desired filters or restrictions.

4. Apply the changes by clicking either the Apply or Save Settings button located at the bottom of the screen.

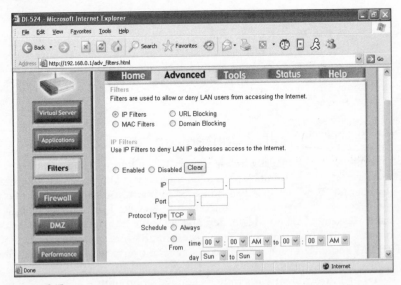

Figure 7-15

8

Adding Peripherals to Your Network

N ow that you have set up your network and fine-tuned its features, it is time to add some helpful or fun peripheral devices like printers, webcams, or video game adapters.

Connect Printers to Your Network

One of the most useful features of a wireless network is its ability to share one printer with multiple computers in the same home or office. In addition, you have the freedom to take your computer almost anywhere in your home or office while printing a document in a separate room. That alone makes it worth the time and money you spent setting up your network. There are three ways to connect a printer to a wireless network: make a direct PC connection, use a wired print server, or use a wireless print server. Each option works equally well.

Make a Direct Connection

In this setup, you hook a computer to your network, and then connect a printer to that computer (see Figure 8-1).

Do It Yourself

Connect printers to your network

Make a direct connection

Install a wired print server

Install a wireless print server

Enable print sharing

Add a shared printer

Connect other peripherals to your network

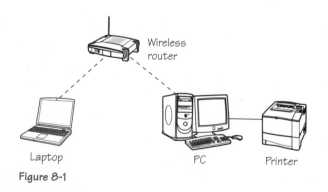

Laptop   PC   Printer

Wireless router

Figure 8-1

Here are some things to consider about direct connections:

- They don't require additional hardware, which is helpful if you are on a tight budget.
- The central computer must be turned on whenever you want to print a document (which is why it is called a "dedicated" computer).
- If the dedicated computer is turned off, no one will be able to print.
- The dedicated computer may suffer slow performance during a print job.
- All documents you want to print must pass through the dedicated computer.

The basic steps to make a direct connection are as follows:

1. Locate your printer's cable (either the USB or 25-pin Centronics style), which should have been included with your printer. If it was not, you need to buy one from a computer or electronics store.

2. Attach one end of the cable to your computer.

3. Attach the other end of the cable to your printer.

4. On the dedicated computer, turn on print sharing. For details on how to do this, see the "Enable Print Sharing" 5-minute fix later in this chapter.

5. For each computer on the network, use Windows to add a shared printer. For details on how to do this, see the "Add a Shared Printer" 5-minute fix later in this chapter.

6. In some cases, you might also need to run a setup CD and install print-server drivers on each computer. Check your printer's instructions to see if you need to do this.

Install a Wired Print Server

In this setup, you connect a wired print server (also referred to as an Ethernet print server) to your router (see Figure 8-2).

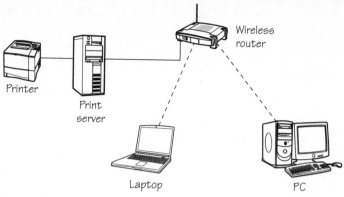

Figure 8-2

Here are some things to consider about using this device:

- It requires additional hardware that usually costs under $100.
- It doesn't require a dedicated computer.
- The printer operates independently and doesn't slow down any computers using the network.
- The printer must be placed near your wireless router (because the printer connects to the print server, which then connects to the router).
- The cables used to connect these devices must be less than 300 feet in length.

The basic steps to install a wired print server are as follows:

1. Purchase an Ethernet print server that has the same interface as the one on your printer (the interface will be serial, parallel, USB, or Ethernet).

Note
Each style of interface uses a different type of cable to link the print server with the printer.

2. Connect the print server to the router, as follows:

   a. Obtain a CAT5 cable long enough to extend from the printer to the router.

   b. Connect one end of the cable into an open Ethernet port on the router.

   c. Connect the other end of the cable into the Ethernet port on the print server.

3. Connect the print server to the printer (which will require a separate cable), as follows:

   a. Using the correct type of cable that matches your printer's interface (serial, parallel, USB, or Ethernet), connect one end of the cable to the print server.

   b. Connect the other end of the cable to the printer.

4. Turn on the print server. It probably won't have a power switch, so just plug it into an electrical outlet.

5. For each computer on the network, use Windows to add a shared printer. For details on how to do this, see the "Add a Shared Printer" 5-minute fix later in this chapter.

6. In some cases, you might also need to run a setup CD and install print-server drivers on each computer. Check your printer's instructions to see if you need to do this.

Install a Wireless Print Server

In this setup, you connect a wireless print server to your printer (see Figure 8-3).

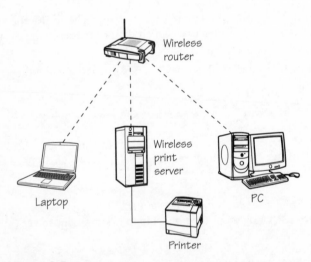

Figure 8-3

Here are some things to consider about using this device:

- It requires additional hardware that can cost more than $100.

- It doesn't require a dedicated computer.

- The printer operates independently and doesn't slow down any computers using the network.

- It allows your printer to be placed almost anywhere in your wireless coverage area.

The basic steps to install a wireless print server are as follows:

1. Purchase a wireless print server that has the same interface as the one on your printer (the interface will be serial, parallel, USB, or Ethernet).

Note
Each style of interface uses a different type of cable to link the print server with the printer.

2. Connect the print server to the printer, as follows:

a. Using the correct type of cable that matches your printer's interface (serial, parallel, USB, or Ethernet), connect one end of the cable to the print server.

b. Connect the other end of the cable to the printer.

3. Turn on the print server. It probably won't have a power switch, so just plug it into an electrical outlet.

4. For each computer on the network, use Windows to add a shared printer. For details on how to do this, see the "Add a Shared Printer" 5-minute fix later in this chapter.

5. In some cases, you might also need to run a setup CD and install print-server drivers on each computer. Check your printer's instructions to see if you need to do this.

Enable Print Sharing

After making a direct connection between your printer and your dedicated computer (as described earlier in this chapter), you must turn on a feature known as Print Sharing. This permits the dedicated computer to share its printer with the rest of your network.

Note
This must be performed on the dedicated computer (which has the printer attached to it).

Follow these steps to turn on Print Sharing with Windows XP Home Edition and XP Professional Edition:

1. Click the Start button in the lower-left corner of Windows.

2. Click the Control Panel. If you don't see this option, your Start menu is in classic mode. In that case, click Settings, and then select the Control Panel.

3. If the Control Panel is in category view, click the Printers and Other Hardware category labeled, and then click the Printers and Faxes labeled. If the Control Panel is in classic view, simply double-click the Printers and Faxes icon.

4. Right-click the printer you want to share, and then click Sharing (see Figure 8-4).

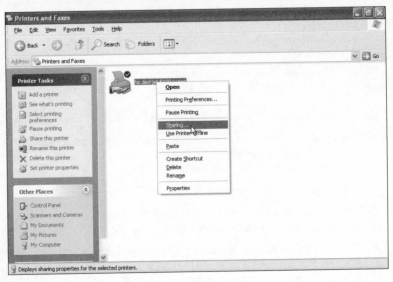

Figure 8-4

5. Click the Share This Printer option (see Figure 8-5).

Figure 8-5

6. Enter a share name for the printer (any name will do), and then click OK.

Follow these steps to turn on Print Sharing with Windows 98 and Windows ME:

1. Click the Start button in the lower-left corner of Windows.

2. Click Settings.

3. Click Control Panel.

4. Click Network.

5. Under the Configuration tab, click File and Print Sharing.

6. Select the option that says "I want to be able to allow others to print to my printer(s)."

7. Click OK.

Add a Shared Printer

After you have connected a printer to your network by making a direct connection, installing a wired print server, or installing a wireless print server, you need to tell each computer on your network where to find the shared printer. This is done by using a Windows feature known as Add Printer.

To use Add Printer with Windows XP Home Edition and XP Professional Edition:

1. Click the Start button in the lower-left corner of Windows.

2. Click the Control Panel. If you don't see this option, your Start menu is in classic mode. In that case, click Settings, and then select the Control Panel.

3. If the Control Panel is in category view, click the Printers and Other Hardware category, and then click the Printers and Faxes icon (previously shown in Figure 8-4). If the Control Panel is in classic view, simply double-click the Printers and Faxes icon labeled.

4. Click the File drop-down menu, and then click Add Printer. The Add Printer wizard appears (see Figure 8-6).

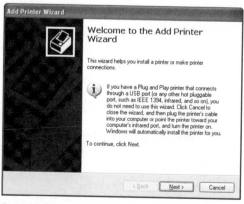

Figure 8-6

5. Click the Next button.

6. Click the A Network Printer, or a Printer Attached to Another Computer option (see Figure 8-7).

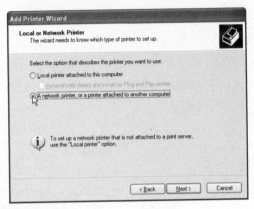

Figure 8-7

7. Click Next.

8. Click the Browse for a Printer option.

9. Find the name of the computer that has the printer attached to it, and then double-click that name.

10. Click the printer you want to add, and then click Next (see Figure 8-8).

FYI

For print sharing to work properly, the dedicated computer (the one that has the printer attached to it) must be running. If you turn off the dedicated computer, you will not be able to print.

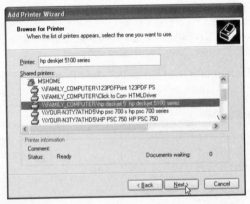

Figure 8-8

11. You are asked if you want to make the printer your default printer (which means it will automatically be used by Windows and your programs to print documents). After making a decision, click Next.

12. Review your information. If it is correct, click Finish.

To use Add Printer with Windows 98 and Windows ME:

1. Click the Start button in the lower-left corner of Windows.

2. Click Settings.

3. Double-click the Printers icon.

4. Double-click the Add Printer icon.

5. Click Next.

6. Select Network Printer, and then click Next.

7. Click Browse, which opens a browsing window. In it, locate your printer, and then click OK.

8. Click Next.

9. Select the brand and model of the shared printer you are adding, and then click Next.

10. Click Next.

11. Choose whether or not you would like to print a test page, and then click Finish.

Connect Other Peripherals to Your Network

In addition to using your computer to send e-mail and surf the Internet, you can increase your fun by adding a wireless video game adapter, a webcam, and a digital media player.

Wireless Video Game Adapters

Wireless game adapters allow you to play interactive, multiplayer, online, or head-to-head video games on popular gaming systems like PlayStation2, Xbox, or GameCube. In this setup, the adapter connects to the video game system and communicates with your wireless router. The router allows the adapter to connect to the Internet or to other wireless video game adapters located nearby. When configuring this adapter, always follow the instructions that came with it. Here is a general overview of the steps required to install one:

1. Buy a wireless video game adapter.

2. Connect a CAT5 cable between the game adapter and your computer.

3. Configure the game adapter by using the setup CD that came with it. You may be required to use your Web browser to configure the adapter.

4. Disconnect the CAT5 cable between the game adapter and your computer.

5. Connect the CAT5 cable between the game adapter and your video game system.

Wireless Webcams

A webcam is a small video camera that connects to your computer and enables you to send still photos or full-motion video to your friends or family over the Internet. It may also have enhanced features like sending automatic e-mails enhanced with audio and video when it senses motion; remote zoom, pan and tilt; night vision; and a weather-proof enclosure for outdoor surveillance. The most recent versions of webcams are wireless, allowing you to place them almost anywhere in your home or office (which is beneficial when using them for security purposes). When configuring a wireless webcam, always follow the instructions that came with it. Here is a general overview of the steps required to install one:

1. Buy a wireless webcam.

2. Connect a CAT5 cable between the webcam and the Ethernet port on your wireless router.

3. Configure the webcam by using the setup CD that came with it.

4. Unplug the webcam's power cord and Ethernet cable from your wireless router.

5. Find a location to mount your webcam.

6. Connect the power cord to the webcam.

> *FYI*
>
> When installing a wireless webcam outdoors, avoid pointing its lens directly east or west. Direct sunlight will significantly distort the video signal and could damage the camera.

Wireless Media Players

A wireless media player allows you to beam music, photos, and videos from your computer to your home stereo system and television. When configuring a wireless media player, always follow the instructions that came with it. Here is a general overview of the steps required to install one:

1. Buy a wireless media player.

2. Most of these media players come with a setup CD containing software, which you should install on your computer.

3. Connect the wireless media player to your television and stereo system by using standard RCA, S-Video, or coaxial cables.

4. Configure the wireless media player by following its manufacturer's user guide.

Internet Telephones

In recent years, a technology called VoIP has seen its popularity surge. VoIP stands for voice over IP. Basically, this means you use the Internet to make phone calls rather than traditional telephone wires. The advantage of VoIP is its price: usually it is significantly lower in cost and offers more features than old-fashioned telephone services. But there are some downsides. If your electricity goes out, your VoIP phone won't work. Also, because a VoIP phone number is portable (meaning it is associated with a computer address, not a physical street address), ambulances and police have had problems determining VoIP users' locations when responding to 911 calls. Fortunately, improvements have been made in this area, so if you want to save money, you might give VoIP some serious thought. Here is a general overview of the steps involved in setting up a VoIP phone:

1. Subscribe to a service plan with a VoIP provider like Vonage. To do so, go to the provider's website or visit a local computer and electronics store.

2. To use your existing telephone with a VoIP service, you need a phone adapter to link your telephone to the Internet. Companies like Linksys make wireless routers that have built-in phone adapters designed to work with Vonage or similar VoIP services. Usually these routers are sold in computer or electronics stores or on their manufacturers' websites.

3. Unplug your standard telephone from its existing telephone jack, and then plug it into one of the phone-adapter ports on the VoIP router. With a traditional wired telephone, you are restricted to using it near the router. But if you have a cordless phone, you can place its base station near the router and use the cordless handset anywhere in your home or office.

4. Most VoIP routers have another phone-adapter port that can be used to hook up a second telephone or a fax machine. In fact, both phone-adapter ports can be assigned different telephone numbers. So if you have a fax or second phone, connect them to the appropriate port on your router.

5. Register your phone-adapter ports with your VoIP provider by visiting its website.

Consider using a special Wi-Fi phone designed to work seamlessly with a VoIP router. This phone sends its signals directly to the VoIP router, which then sends them over the Internet to their destination. Not only can you use a Wi-Fi phone at your home or office, but you also can take it on the road and use it at public Wi-Fi hotspots like the ones found in many popular coffee shops or bookstores.

9

Maintaining Your Network

After you have set up, configured, and tweaked your wireless network, there are still some things you must periodically do to keep it running at an optimal level or to improve its performance:

- **Change the data rate:** To make your wireless router's signal travel farther, you can decrease (slow down) its data rate. On the flip side, you can make your wireless network faster by increasing the data rate, but this will cause the signal to travel shorter distances (which means you will have to move closer to your router when using a wireless computer).

- **Reduce interference:** If there are other wireless networks nearby, use a different channel than theirs (for more information, see Chapter 10).

- **Don't use certain types of cordless phones:** Cordless phones that operate in the 2.4 GHz frequency can interfere with wireless networks that use the 802.11b or 802.11g technologies (both of which also operate in the 2.4 GHz frequency).

- **Move your router away from microwave ovens:** When a microwave oven is operating, it can interfere with your wireless network. Therefore, do not place your router anywhere near a microwave.

- **Use 802.11g:** If you are using the older 802.11b technology, consider upgrading to 802.11g to give your network a boost of speed. This extra speed doesn't necessarily make the Internet run faster, but it greatly improves the rate at which your files travel across your network.

Do It Yourself

Increase the performance of your network

Change the data rate

Change the router's channel

Upgrade to 802.11g

Extend the range of your network

Change your router's transmit power

Upgrade your network

Upgrade windows XP to support WPA

Upgrade your router's firmware

Upgrade your wireless card

Increase the Performance of Your Network

When using your wireless network, you may find that its default settings don't deliver the kind of performance you need. In that case, try one of these 5-minute fixes.

Change the Data Rate

In most cases, you should let your wireless cards and router automatically select the speed at which your information is transmitted across your network (referred to as the *data rate*). The only time you should change the default settings for the data rate is when you want your router's signals to travel longer or shorter distances. For example, if your current signal cannot reach a particular room in your home or office, you can lower the data rate in order to make the signal extend into that room.

Follow these steps to change the data rate for wireless routers:

1. Log in to your wireless router's configuration utility, as follows:

   a. Make sure you are connected to your wireless network. Test this by hovering your mouse pointer over the wireless icon located in the lower-right corner of Windows. A message indicates your connected status.

   b. Open a Web browser like Internet Explorer. If your router is not connected to the Internet, your browser will not find a valid Web page. Instead, it displays an error message that says something like "Page cannot be displayed" or "Page not found."

   c. In your Web browser's address window (the place where you usually type the name of a website you want to visit), type your router's IP address, and then press the Enter key. Do not type http:// or www. Instead, only type the number for your IP address.

   d. A login box appears. If you created a username and password when initially setting up your router, type them into the login box. If you didn't create a username or password, look for the default ones listed in your wireless router's instructions or manual.

2. Locate the Advanced Wireless (or Performance) section, and then find the Data Rate setting (which some routers call Rate, TxRate, or Basic Rate). Figure 9-1 shows a typical data-rate screen.

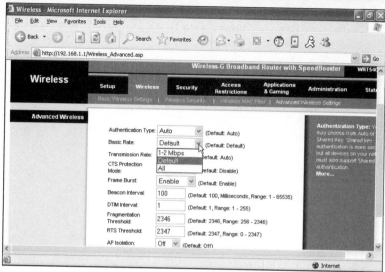

Figure 9-1

3. Choose the desired data-rate setting, following these guidelines:

   a. "Auto" means the router automatically determines the best data rate to use.

   b. Changing the data rate to a very low value, such as 1 Mbps, makes your wireless signal travel farther.

   c. Choosing a high fixed data rate, such as 54 Mbps, limits the range of your network but makes it safer from high-tech criminals. Be advised that this high setting could make it difficult for computers to use your network if they are located too far away from your router.

4. Save your changes by clicking either the Apply or Save Settings button located at the bottom of the screen.

Follow these steps to change the data rate for wireless cards:

1. Double-click the wireless icon located in the lower-right corner of Windows.

2. Click Properties.

3. Click Configure.

4. Click the Advanced tab. If this tab isn't available, then unfortunately you cannot change the data rate by using Windows. Instead, you have to use the manufacturer's configuration utility that came with your wireless card (see Chapter 5 for more information on using the manufacturer's utility).

5. Find the Data Rate settings (which some wireless card call them Rate, TxRate, or Basic or Transmission Rate). Figure 9-2 shows a typical data-rate screen for a wireless card.

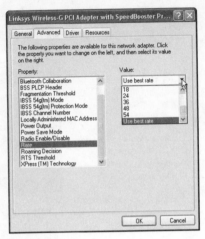

Figure 9-2

6. Select the desired data rate.

Change the Router's Channel

If your wireless router is having problems sending or receiving signals, it may be due to interference from nearby routers used by your neighbors. One solution is to change your router's channel. Follow these steps:

Note
You only need to change the channel in your wireless router because your wireless cards automatically tune to the channel being used by the router.

1. Log in to your wireless router's configuration utility, as follows:

a. Make sure you are connected to your wireless network. Test this by hovering your mouse pointer over the wireless icon located in the lower-right corner of Windows. A message should indicate your connected status.

b. Open a Web browser like Internet Explorer. If your router is not connected to the Internet, your browser will not find a valid Web page. Instead, it displays an error message that says something like "Page cannot be displayed" or "Page not found."

c. In your Web browser's address window (the place where you usually type the name of a website you want to visit), type your router's IP address, and then press the Enter key. Do not type http:// or www. Instead, only type the number for your IP address.

d. A login box appears. If you created a username and password when initially setting up your router type them into the login box. If you didn't create a username or password, look for the default ones listed in your wireless router's instructions or manual.

2. Find the configuration screen containing the settings for the router's channel. Figure 9-3 shows a typical channel screen.

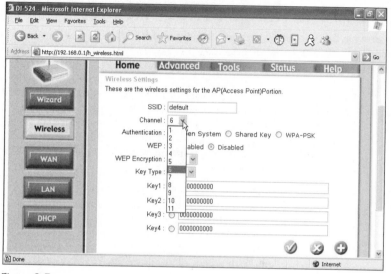

Figure 9-3

3. Change the channel to 1 or 11.

4. Save your changes by clicking either the Apply or Save Settings button located at the bottom of the screen.

Upgrade to 802.11g

If you have an older wireless router that uses the 802.11b technology, you might be able to upgrade it to 802.11g (which would boost the speed of your wireless network from 11Mbps to 54Mbps). Here's how:

Note

Not all routers are capable of being upgraded to 802.11g.

1. Check the website belonging to the router's manufacturer to see if there are any details about upgrading to 802.11g. This information is usually under the Support or Downloads section. If you can't find the answers on the website, contact the manufacturer's customer-service department.

2. If an upgrade is available, typically it comes in the form of a firmware patch that must be downloaded and applied to your router. Follow the directions on the manufacturer's website to download this patch.

3. To install the patch, carefully read the instructions that came with it (or perhaps they are available on the manufacturer's website).

4. After the patch is installed, you are instructed to reboot the router. You can do this by unplugging the router's electrical cord from the wall outlet, waiting 15 seconds, and then plugging it back in.

Extend the Range of Your Network

To provide full wireless coverage throughout a large home or office, you might need to extend the range of your network. Here are some general guidelines to follow:

- **Increase the transmit power:** In most wireless routers and cards, the default transmit power is already set at the maximum level. However, you should double-check these settings before spending money on extra devices to extend the range of the network. For more information on accessing the transmit power settings, see the corresponding 5-minute fix in this chapter.

- **Use better antennas:** If your wireless router has a removable antenna, you can take it off and buy a better, stronger model that will make your wireless signals travel farther. Usually you can purchase these antennas at a computer or electronics store or on the website belonging to the manufacturer of your wireless router. Be sure to choose an antenna with a gain of at least 6dB.

- **Use an amplifier:** To increase the overall power of your wireless network and allow computers to access your router from longer distances, you can connect a special amplifier between your router and your antenna. But be warned: typically it is illegal to use an amplifier without getting a special license from the Federal Communications Commission (the FCC).

- **Use a repeater:** A device known as a repeater allows you to make your wireless network reach longer distances. Ideally you should place the repeater at the point where your wireless signal begins to fade and grow weak. In particular, a repeater is a great tool for extending an indoor wireless network to your backyard.

Change Your Router's Transmit Power

If you suspect that your router might not be transmitting wireless signals at the highest possible level of power, check its settings by doing the following:

1. Log in to your wireless router's configuration utility as follows:

   a. Make sure you are connected to your wireless network. Test this by hovering your mouse pointer over the wireless icon located in the lower-right corner of Windows. A message indicates your connected status.

b. Open a Web browser like Internet Explorer. If your router is not connected to the Internet, your browser will not find a valid Web page. Instead, it displays an error message that says something like "Page cannot be displayed" or "Page not found."

c. In your Web browser's address window (the place where you usually type the name of a website you want to visit), type your router's IP address, and then press the Enter key. Do not type http:// or www. Instead, only type the number for your IP address.

d. A login box appears. If you created a username and password when initially setting up your router, type them into the login box. If you didn't create a username or password, look for the default ones listed in your wireless router's instructions or manual.

2. Find the configuration screen containing the settings for the router's transmit power (see Figure 9-4).

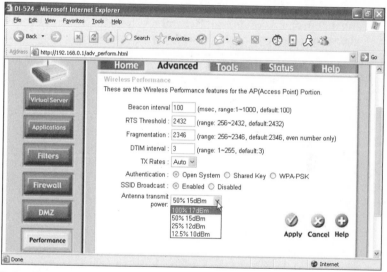

Figure 9-4

3. Change the transmit power level to the highest setting (100 percent, highest, or 100mW).

4. To improve your router's security, you can change its transmit power to a lower level (which also reduces the distance your wireless network can reach).

Upgrade Your Network

Many hardware or software items in your wireless network are capable of being upgraded with patches or enhancements offered by their manufacturers. Keeping your wireless components up-to-date is a simple way to give them a long, productive lifespan.

Upgrade Windows XP to Support WPA

WPA is a high-level encryption mode for wireless networks that provides outstanding protection against high-tech criminals. If you are using Windows XP without Service Pack 2 and want to use WPA, you will probably need to download a free update from Microsoft's website. This update enables you to use the WPA encryption methods through the Windows XP wireless configuration utility. If you aren't sure whether you have Service Pack 2 installed on your Window XP computer, here's how you can check:

1. Click the Start button in the lower-left corner of Windows.

2. Click the Control Panel. If you don't see this option, your Start menu is in classic mode. In that case, click Settings, and then select the Control Panel.

3. If the Control Panel is in category view, click the Performance and Maintenance category, and then click the System icon. If the Control Panel is in classic view, simply double-click the System icon.

4. A window opens. Under the General tab, look for System. Beneath it, you should see some words identifying your version of Windows XP as well as any service packs that are installed (see Figure 9-5).

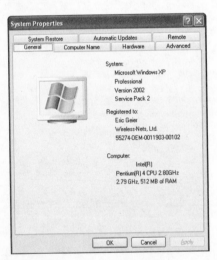

Figure 9-5

5. If you already have Service Pack 2 but can't get WPA encryption to work properly, read the instructions that came with your wireless card to see if WPA is actually supported by your card.

6. If you do not have Service Pack 2, you need to download the WPA update from Microsoft by doing the following:

    a. Connect to the Internet and visit support.microsoft.com/ kb/826942 (which is Microsoft's Support website).

    b. Scroll down to the More Information section.

c. Click the download link that corresponds to your version of Windows XP (either 32-Bit or 64-Bit) — see Figure 9-6.

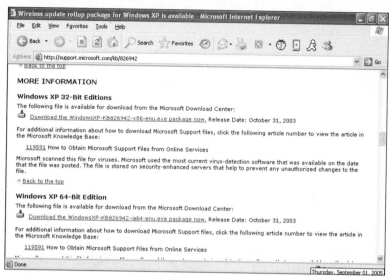

Figure 9-6

d. Click Download.

e. Click Save, select a location on your computer where you want to save the file, and then click OK.

f. After the download is complete, click either the Open or Run button on the download window. Follow the on-screen instructions to install the update. If you see a security warning, click the Run button next to it.

Upgrade Your Router's Firmware

The brain of a router is its *firmware*, which is a set of instructions that controls its activities. Occasionally manufacturers release new firmware to offer enhanced features, fix errors, or plug security holes. To make sure your router is safe and secure, you should check for new firmware periodically. Here's how:

1. Log in to your wireless router's configuration utility, as follows:

a. Make sure you are connected to your wireless network. Test this by hovering your mouse pointer over the wireless icon located in the lower-right corner of Windows. A message indicates your connected status.

b. Open a Web browser like Internet Explorer. If your router is not connected to the Internet, your browser will not find a valid Web page. Instead, it displays an error message that says something like "Page cannot be displayed" or "Page not found."

   c. In your Web browser's address window (the place where you
      usually type the name of a website you want to visit), type your
      router's IP address, and then press the Enter key. Do not type
      http:// or www. Instead, only type the number for your IP
      address.

   d. A login box appears. If you created a username and password
      when initially setting up your router, type them into the login
      box. If you didn't create a username or password, look for the
      default ones listed in your wireless router's instructions or
      manual.

2. Find the configuration screen containing information about the
   firmware. This is located under either the Router Information or
   Status section (see Figure 9-7).

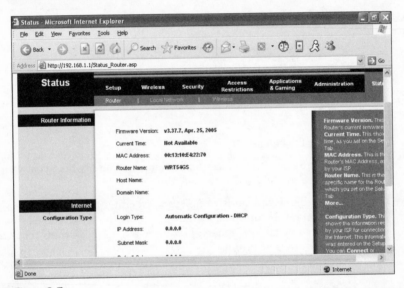

Figure 9-7

3. Grab a pen and paper and write down your firmware's version and
   release date.

4. Connect to the Internet and visit the website belonging to your
   router's manufacturer.

5. Click the Support or Downloads section of the website, and then look
   for information about upgrading firmware (see Figure 9-8).

5. Find the make and model of your wireless router.

6. Compare the version number of your current firmware with the one
   available on the website. If the website's firmware has a higher number
   than yours, that means your current firmware is outdated. Go ahead
   and download the new one. Be sure to save it in a folder you can easily find, because you will need to access that folder when you update
   the firmware in the next step.

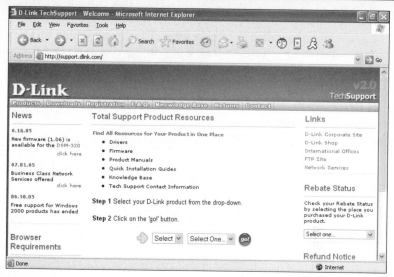

Figure 9-8

7. Return to your router's configuration screen by following the directions in Step 1.

8. Find the Upgrade Firmware (or similarly named) setting. Often it is located under a Tools or Administration section. Figure 9-9 shows a typical firmware-update page.

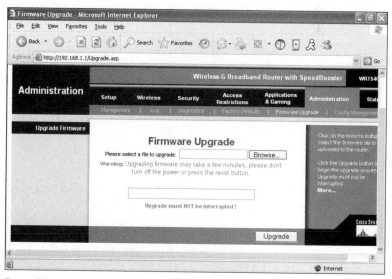

Figure 9-9

9. Click Browse, which opens a browsing window. Use it to locate the new firmware you just downloaded. When you locate the firmware, double-click it.

10. Click OK, and then follow the on-screen instructions.

Upgrade Your Wireless Card

To get the best performance from your wireless card, you should update it with the most recent *drivers*, which are a type of software that controls your card. Occasionally manufacturers release new drivers for their products to offer enhanced features, fix errors, and plug security holes.

Determine the Current Version

Before making changes to your drivers, you must find out what version they are.

To find the driver version on Windows XP Home Edition and XP Professional Edition:

1. Double-click the wireless icon located in the lower-right corner of Windows.

2. Select Properties.

3. Click Configure.

4. Click the Driver tab (see Figure 9-10).

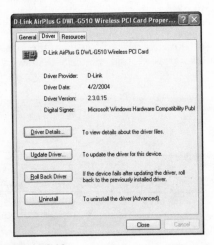

Figure 9-10

5. Grab a pen and paper and jot down the version number and release date for your wireless card's drivers.

To find the driver version on Windows 98:

1. Right-click the My Computer icon on your desktop.

2. Select Properties.

3. Click the Device Manager tab.

4. Click the plus sign located next to the Network Adapters category.

5. Double-click the wireless card you want to check.

6. Click the Driver tab.

7. Click Driver File Details (see Figure 9-11).

Figure 9-11

8. Grab a pen and paper and jot down the version number and release date for your wireless card's drivers.

Check for New Drivers

To determine if your drivers need to be updated, do the following:

1. Visit the website belonging to the manufacturer of your wireless card. Click on the Support or Downloads section of the site, and then look for information about drivers.

2. Locate the drivers on the website that correspond to your specific brand and model of wireless card.

3. Compare the version number of your current drivers with the ones available on the website. If the website's drivers have a higher number than yours, that means your current drivers are outdated. Go ahead and download the new ones. Be sure to save them in a folder you can easily find later, because you will need to access that folder when you update the drivers in the next step.

Update the Drivers

After downloading the new drivers, you will need to update your wireless card through Windows.

To update the drivers with Windows XP Home Edition and XP Professional Edition:

1. Double-click the wireless icon located in the lower-right corner of Windows.

2. Select Properties.

3. Click Configure.

4. Click the Driver tab.

5. Click Update Driver (see Figure 9-12).

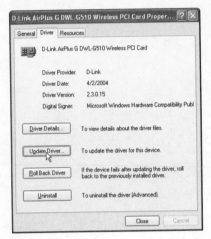

Figure 9-12

6. Select the Install from a List or Specific Location option (see Figure 9-13).

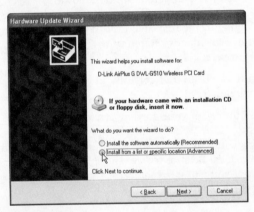

Figure 9-13

7. Click Next.

8. Click the Don't Search, I Will Choose the Driver to Install option.

9. Click Next.

10. Click the Have Disk button (see Figure 9-14).

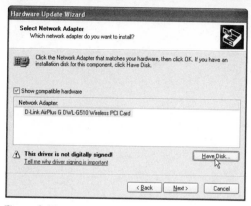

Figure 9-14

11. Locate the drivers you downloaded from the manufacturer's website.

12. To install the drivers, click the Finish button.

To update the drivers with Windows 98:

1. Right-click the My Computer icon on your desktop.

2. Select Properties.

3. Click the Device Manager tab.

4. Click the plus sign located next to the Network Adapters category.

5. Double-click the wireless card you want to update.

6. Click the Driver tab.

7. Click Update Driver.

8. Follow the on-screen instructions.

PART V

TROUBLESHOOTING

Hopefully, nothing will go wrong with your wireless network. But if it does, turn to this section of the book for information on troubleshooting your wireless problems and solving common issues.

10

Troubleshooting

Like any computer equipment, a wireless network is bound to have problems that will make you want to tear your hair out. Fortunately you can avoid going bald by following the troubleshooting techniques in this chapter.

Network Connection Problems

Probably the most common problem you will have with your wireless network is difficulty connecting to it or maintaining a steady connection. To fix these issues, there are several 5-minute fixes you can try.

Reboot Your Router

A quick and easy way to solve many connection problems is to reboot your router, like this:

1. Make sure your wireless router is plugged into an electrical outlet and is powered on properly. This is indicated by the blinking lights on the front of the router.

2. Reboot both the router and the broadband modem by unplugging their electrical cords from the wall outlet, then waiting 15 seconds before plugging them back in. If this doesn't solve the problem, move on to the next step.

3. Try disabling and then re-enabling the wireless card in your computer by doing the following:

   a. Right-click the wireless icon in the lower-left corner of Windows.

   b. Click Disable.

   c. Click the Start button in the lower-left corner of Windows.

    d. Choose Connect To.

    e. Click Wireless Network Connection.

4. If you are still having problems, your network might be receiving interference. To reduce this interference, see the next 5-minute fix.

Improve a Poor Connection

If your wireless signal fades in and out, it is probably the result of interference or a weak signal.

To reduce interference do not place a microwave oven or a cordless telephone using the 2.4 GHz frequency within 10 feet of your wireless network.

To improve a weak signal, try decreasing the distance between the wireless card in your computer and the wireless router. If you are using a desktop computer, try putting the PC case on your desk to give the wireless card's antenna a better vantage point for receiving the signal.

When Wireless Computers Won't Communicate

If one computer on your wireless network cannot access the shared folders or printers on another computer, try this remedy:

1. Check to see if each wireless computer is connected to the same network. This can be done by hovering your mouse pointer over the wireless icon located in the lower-right corner of Windows. The network name appears next to Wireless Network Connection (see Figure 10-1). If this name is not the same on all of your computers, then connect each computer to the same network.

2. Restart the wireless computers and wait a few minutes.

Figure 10-1

Use Wireless Network Analyzers

A *network analyzer* is a device that shows you specific details about your wireless network as well as other networks located nearby. By knowing what wireless settings your neighbors are using, you can change yours to avoid conflicts. Here are some popular analyzers (in no particular order):

- Network Stumbler (see Figure 10-2)

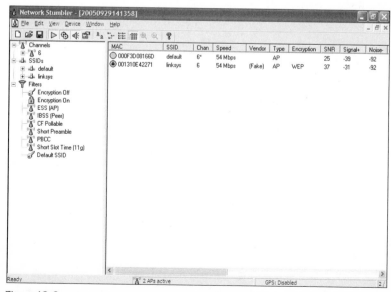

Figure 10-2

- AirMagnet (see Figures 10-3 and 10-4)

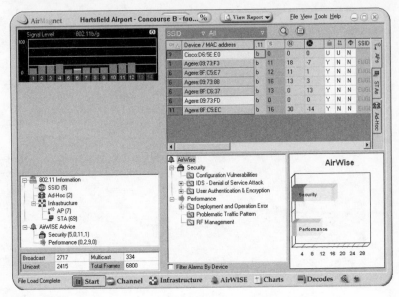

Figure 10-3

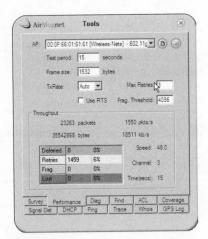

Figure 10-4

- WilPackets
- Ekahau
- Helium Networks

Most of these analyzers display information like:

- **SSIDs:** This is useful to identify any other wireless networks nearby that may be interfering with yours.

- **RF Channels:** Find the channels in use by neighboring wireless networks, and then set yours to a nonconflicting one.

- **Speed:** This tells you what data rates the wireless devices are using.

- **Signal:** This tells you how strong or weak your wireless signals are. These values will generally be negative numbers. Keep in mind that smaller negative numbers actually mean a higher signal level.

- **Noise:** This is the amount of interference caused by weather conditions outside or by man-made devices like microwave ovens. Under normal conditions, without any significant inference, this value will generally be less than negative 90dBm. Higher values (smaller negative numbers) suggest that interference might be present.

Internet Connection Problems

If the computers on your wireless network cannot access the Internet, try one of these fixes:

- Turn off and restart your broadband modem, wireless router, and computer. This gives everything a fresh start. To restart your broadband/DSL modem and wireless router, unplug their power cords, wait 15 seconds, and then plug them in again.

- Make sure your broadband modem is turned on. When it is plugged in properly, the lights on its front should illuminate (and some even blink). If those lights are on but you still cannot access the Internet, contact the company that provides your high-speed Internet.

Enable RTS/CTS

If you have multiple users on your wireless network and all of them are experiencing connection difficulties, it could be the result of their computers sending signals that collide with each other. To solve this problem, turn on a feature known as RTS/CTS. Here's how.

Note
Some wireless cards don't allow you to use RTS/CTS.

1. Double-click the wireless icon in the lower-right corner of Windows.
2. Click Properties.
3. Click Configure.

4. Click the Advanced tab (see Figure 10-5). If this tab isn't available, then unfortunately you can't turn on RTS/CTS through Windows. Instead, you have to use the manufacturer's configuration utility that came with your wireless card. For more information on accessing the manufacturer's utility, refer to Chapter 5.

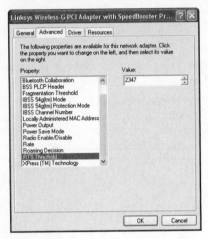

Figure 10-5

5. Find the RTS/CTS settings.

6. Set the RTS/CTS threshold to 750.

Enable Fragmentation

Another way to counteract poor performance on your network is to turn on a feature called *fragmentation* in both your wireless card and router. Here's how.

1. Log in to your wireless router's configuration utility as follows:

  a. Make sure you are connected to your wireless network. Test this by hovering your mouse pointer over the wireless icon located in the lower-right corner of Windows. A message indicates your connected status.

  b. Open a Web browser like Internet Explorer. If your router is not connected to the Internet, your browser will not find a valid Web page. Instead, it displays an error message that says something like "Page cannot be displayed" or "Page not found."

  c. In your Web browser's address window (the place where you usually type the name of a website you want to visit), type your router's IP address, and then press the Enter key. Do not type http:// or www. Instead, only type the number for your IP address.

d. A login box appears. If you created a username and password when initially setting up your router, type them into the login box. If you didn't create a username or password, look for the default ones listed in your wireless router's instructions or manual.

2. Find the configuration screen containing the fragmentation settings. They are usually in the Advanced Wireless or Performance section.

3. Set the fragmentation threshold to 750 (see Figure 10-6).

Figure 10-6

4. Save the changes by clicking either the Apply or Save Settings button located at the bottom of the screen.

Enable Protection Mechanisms

If you have a large wireless network with many users, consider turning on the Protection Mechanisms. This will prevent computers using the 802.11b wireless cards from interfering with computers using the 802.11g cards.

Note
Some wireless routers don't allow you to use protection mechanisms.

To turn on the Protection Mechanisms:

1. Log in to your wireless router's configuration utility, as follows:

a. Make sure you are connected to your wireless network. Test this by hovering your mouse pointer over the wireless icon located in the lower-right corner of Windows. A message indicates your connected status.

b. Open a Web browser like Internet Explorer. If your router is not connected to the Internet, your browser will not find a valid Web page. Instead, it displays an error message that says something like "Page cannot be displayed" or "Page not found."

c. In your Web browser's address window (the place where you usually type the name of a website you want to visit), type your router's IP address, and then press the Enter key. Do not type http:// or www. Instead, only type the number for your IP address.

d. A login box appears. If you created a username and password when initially setting up your router, type them into the login box. If you didn't create a username or password, look for the default ones listed in your wireless router's instructions or manual.

2. Find the configuration screen that contains the settings for protection mechanisms (which is sometimes referred to as CTS Protection Mode). It is usually located in the Advanced Wireless or Performance section.

3. Choose your desired setting: Disable or Auto (see Figure 10-7).

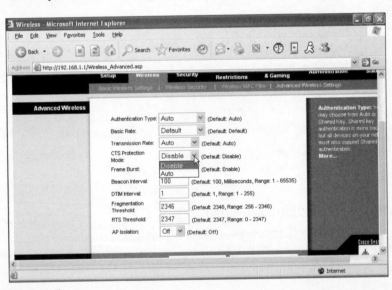

Figure 10-7

4. Save the changes by clicking the Apply or Save Settings button located at the bottom of the screen.

Enable 802.11g Mode Only

If you have an 802.11g wireless router, you can set your router to *802.11g-only* mode, which prevents stray 802.11b users (such as your neighbors) from connecting to your network and slowing its performance. Here's how:

1. Log in to your wireless router's configuration utility, as follows:

   a. Make sure you are connected to your wireless network. Test this by hovering your mouse pointer over the wireless icon located in the lower-right corner of Windows. A message indicates your connected status.

   b. Open a Web browser like Internet Explorer. If your router is not connected to the Internet, your browser will not find a valid Web page. Instead, it displays an error message that says something like "Page cannot be displayed" or "Page not found."

   c. In your Web browser's address window (the place where you usually type the name of a website you want to visit), type your router's IP address, and then press the Enter key. Do not type http:// or www. Instead, only type the number for your IP address.

   d. A login box appears. If you created a username and password when initially setting up your router, type them into the login box. If you didn't create a username or password, look for the default ones listed in your wireless router's instructions or manual.

2. Find the configuration screen containing the 802.11g-only mode (which is also referred to as Wireless Mode). You can usually find this setting in the Basic Wireless section labeled.

3. Select the setting you want. If you enable G-Only (802.11g-only), then 802.11b wireless users will not be able to connect to the router. If you enable B-Only (802.11b-only), then 802.11g wireless users will not be able to connect to the router. If you enable Mixed (also called "b and g"), then both 802.11g and 802.11b wireless users will be able to connect to the router. See Figure 10-8.

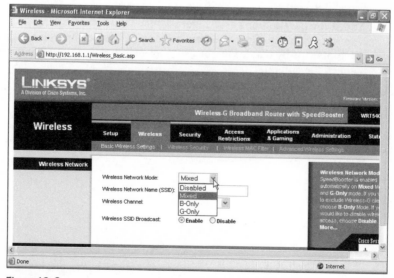

Figure 10-8

4. Save the changes by clicking the Apply or Save Settings button located at the bottom of the screen.

Antenna Problems

Another potential cause of your connection problems is a missing or damaged antenna.

To check for missing antennas:

1. See if any of the antennas are missing on your wireless card or router. It is possible that one of them was accidentally snapped off, or perhaps you forgot to install an antenna when initially setting up your router or card (which has been known to happen to many computer users).

2. Make sure that all of the antennas are connected tightly. If an antenna has come loose, it will not properly send or receive wireless signals.

3. Inspect the entire antenna to look for cracks. Their presence might indicate internal damage to the antenna. If you suspect this to be true, replace the antenna.

If you need to replace a damaged antenna, or if you want to install a more powerful antenna you purchased at a computer or electronics store, then do the following:

1. Turn off the power to the wireless computer or wireless router and unplug it from the electrical outlet. This protects you from receiving an electrical shock.

2. Remove the existing antenna from the wireless card or router, which you can usually do by unscrewing it.

3. Attach the new antenna (see Figure 10-9). First, insert the tip of the antenna's connector into the hole on the wireless device. Next, tighten the connector by hand. Do not use a wrench, because doing so could damage the antenna.

4. Turn on the wireless computer or router by plugging it into the electrical outlet.

5. Make sure the wireless signals are being properly sent and received. You should have steady coverage throughout most of your home or office.

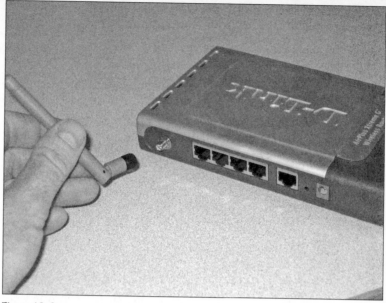

Figure 10-9

Find a Lost or Forgotten Encryption Key

If your wireless network is secured using the WEP or WPA-PSK encryption modes, you must type an encryption key (similar to a password) whenever you attempt to access your network. If you lose or forget this key, try one of these remedies.

Access the Router through another Computer

Assuming you have another wireless computer that has been configured with the correct security settings, use it to access the router's configuration utility, where you can *possibly* find your encryption key. However, be warned that for security reasons, the encryption keys may not be visible in the router. If that is the case with your router, you will have to reset its factory-default settings by following the next 5-minute fix.

Reset the Router's Default Settings

There are two ways to restore a wireless router to its factory-default settings: press the reset button, or access the router's configuration utility.

Note

Restoring your wireless router back to its factory-default settings will wipe out all of your settings, including encryption keys, passwords, and Internet settings.

To use the reset button to restore the router's default settings:

1. Look for a tiny button located on the back of the router. This button may be labeled Reset or something similar. Don't worry if you can't find this button, because some routers don't have reset buttons.

2. Hold down the reset button for at least 10 seconds. Due to the small size of the button, you may need to use a pencil or paperclip to do this.

3. Release the button. The router should reboot, causing the lights on the front of the router to go dark then come back on again. It may take a couple of minutes before the router is usable again.

4. Because the restore operation reverts the router to its factory-default settings, you need to reconfigure the router with the appropriate SSID, channel setting, and security. For more information on this process, refer to the 5-minute fixes in Chapter 4.

To use the Configuration Utility to restore the router's default settings:

1. Log in to your wireless router's configuration utility, as follows:

   a. Make sure you are connected to your wireless network. Test this by hovering your mouse pointer over the wireless icon located in the lower-right corner of Windows. A message indicates your connected status.

   b. Open a Web browser like Internet Explorer. If your router is not connected to the Internet, your browser will not find a valid Web page. Instead, it displays an error message that says something like "Page cannot be displayed" or "Page not found."

   c. In your Web browser's address window (the place where you usually type the name of a website you want to visit), type your router's IP address, and then press the Enter key on your keyboard. Do not type http:// or www. Instead, only type the number for your IP address.

   d. A login box appears. If you created a username and password when initially setting up your router, type them into the login box. If you didn't create a username or password, look for the default ones listed in your wireless router's instructions or manual.

2. Find the restore feature, which may be located in the System, Administration, Utilities, or Maintenance section.

3. Press or select the Restore To Factory Default Settings option. Figure 10-10 shows a typical restore screen.

4. The router should reboot, causing the lights on the front of the router to go dark then come back on again. It may take a couple of minutes before the router is usable again.

5. Reconfigure the router (for details on how to do this, refer to Chapter 4). Because the restore operation reverts the router back to its factory-default settings, you need to reconfigure the router with the appropriate SSID, channel setting, and security.

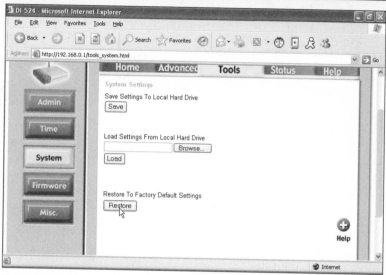

Figure 10-10

Find a Lost or Forgotten User Name or Password

If you have lost or forgotten the user name or password that you are required to type in order to access your router's configuration utility, you need to restore the router's factory-default settings by following the steps in the previous 5-minute fix.

GLOSSARY

These terms can help you through the process of purchasing and installing your wireless network:

802.11: This is the official standard for wireless local area networks (LANs), which includes interactions between the wireless adapters in computers, routers, and access points. The Institute of Electrical and Electronic Engineers (IEEE) is the organization that ratifies the 802.11 standard. 802.11 provides the basis for Wi-Fi.

802.11a: This version of the 802.11 standard specifies data rates up to 54 Mbps in the 5 GHz band using orthogonal frequency division multiplexing (OFDM) technology. 802.11a offers very high capacity, but it is not compatible with any of the 2.4 GHz standards, such as 802.11b and 802.11g.

802.11b: This version of the 802.11 standard specifies data rates up to 11 Mbps in the 2.4 GHz band using direct sequence spread spectrum technology. 802.11b is compatible with 802.11g, but not 802.11a. For example, a computer having an 802.11b wireless adapter will interface with an 802.11g router or access point.

802.11g: This version of the 802.11 standard specifies data rates up to 54 Mbps in the 2.4 GHz band using orthogonal frequency division multiplexing (OFDM). Despite its higher data rates, 802.11g offers lower capacity than 802.11a. 802.11g, however, is backward-compatible with 802.11b wireless adapters.

ad hoc network: This is an optional mode of operation for 802.11 (and Wi-Fi) wireless devices. When you specify an ad hoc network, the wireless adapters in each of the computers (with ad hoc mode enabled), will communicate wirelessly directly with each other without the need for a wireless access point or router.

AES: The Advanced Encryption Standard (AES) is the strongest encryption that 802.11 wireless LANs implement. Not all existing wireless routers and access points can support AES encryption because of special hardware requirements.

broadband modem: This hardware device interfaces a wireless router to the Internet. You must subscribe to an Internet service provider (ISP) before accessing the Internet.

category 5 cable: This cable has unshielded twisted pair wires enclosed in a yellow or blue casing. You need Category ("Cat") 5 cabling to interconnect devices via Ethernet, such as when connecting a wireless router to a broadband modem.

DHCP: Dynamic Host Configuration Protocol (DHCP) automatically assigns unique Internet Protocol (IP) addresses to network devices, such as wireless adapters and wireless routers. In most cases for home and small office applications, it's best to enable DHCP on all wireless adapters. The DHCP server located in the wireless router then automatically assigns an applicable IP address to each wireless adapter. This is also necessary when you are connecting to public wireless hotspots.

diversity: This is a technique whereby two antennas, spaced apart by a specific distance, on the wireless router or access point strengthen the reception of wireless signals. This improves range of the wireless network.

Ethernet: The industry name of the standard for local area networks that uses cabling between computers and networks. The official standard that Ethernet applies to is IEEE 802.3.

fragmentation: This is an optional 802.11 (and Wi-Fi) protocol that breaks apart 802.11 data frames into smaller pieces that are sent separately between wireless adapters and routers. Fragmentation can improve performance in the presence of radio frequency interference.

hotspot: This is a place where you can access the Internet using the wireless adapter in your computer. Some hotspots are free, others charge a fee.

MIMO: The multiple input multiple output (MIMO) technology defines techniques for smart antennas that can improve the performance of a wireless network.

PoE: Power-over-Ethernet (PoE) is a technology that allows you to provide electrical power to a wireless router or access point over the Category 5 Ethernet cable that connects the wireless router or access point to the wired network. Thus, PoE eliminates the need to find or install electrical power outlets near the installation location of the wireless adapter or access point.

power-save mode: When you switch a wireless adapter to power-save mode, the adapter enters "sleep" state when it does not need to send data. This may conserve the battery in your wireless device.

RTS/CTS: Request to send/clear to send (RTS/CTS) is an optional mechanism that you can set on wireless adapters and wireless routers and access points. RTS/CTS set on the wireless adapters may improve performance if the adapters are connecting to the same router (but the adapters are out of range of each other).

SSID: The service set identifier (SSID) is the name that you give your 802.11 (or Wi-Fi) wireless network. You configure this value in the wireless router/access point. Unless you disable SSID broadcasting, the SSID value is what shows up as the network name when users view available wireless networks. If you disable SSID broadcasting, you must manually configure the SSID in each wireless adapter.

SSL: The Secured Socket Layer (SSL) protocol offers end-to-end encryption between wireless computers and severs, such as websites. You should ensure that SSL is in use (indicated by the small pad lock symbol on the screen) when you purchase items over an Internet connection.

VoIP: VoIP (voice over IP) technologies enable voice communications over both wired and wireless networks.

WEP: The Wired Equivalent Protoocl (WEP) is the initial 802.11 security technique. It's possible for hackers to crack WEP encrypted communications, so only use it if nothing else is available. Instead of WEP, use at least TKIP (same as WPA) or AES (same as WPA2).

Wi-Fi: This is the body of standards based on the 802.11 that ensures interoperability among different manufacturers of wireless network adapters and routers/access points. The Wi-Fi Alliance requires the manufacturer of each Wi-Fi certified product to meet specific test requirements.

Wi-Fi zone: This is a public hotspot place where you can interface with the Internet using a Wi-Fi wireless adapter.

wireless adapter: This hardware device interfaces computers, such as desktop PCs and notebook computers, to a wireless network.

wireless router: This hardware device includes a wireless access point and other networking protocols, such as DHCP and NAT.

WPA: The Wi-Fi Protected Access (WPA) protocol uses the TKIP protocol to automatically assign (and periodically rotate) encryption keys to wireless adapters. A newer version of WPA (WPA2) makes use of AES.

INDEX

Don't Stop Now . . .

CHECK OUT THE GEMS YOU'LL FIND IN OTHER *GEEKS ON CALL 5-MINUTE FIXES* TITLES

The following pages contain valuable tips and helpful techniques you can use to work with your PC's hardware and software, enhance security and protect your privacy, and customize your Windows XP experience. And this is just a small sampling of the expert advice you get in *Geeks On Call PCs: 5-Minute Fixes, Geeks On Call Security and Privacy: 5-Minute Fixes, and Geeks On Call Windows XP: 5-Minute Fixes.*

Determine the Manufacturer and Speed of Your Processor (CPU)

The processor—also known as a CPU—is the part of your computer's hardware that performs the tasks handed out by your software. Typically, the faster a processor is, the more powerful your computer will be. Some programs and newer versions of Windows require a minimum CPU speed in order to work correctly. To find out how fast your CPU is and which company made it, do the following:

1. Right-click the My Computer icon on your desktop. If this icon is not available, click the Start button in the lower-left corner of Windows and right-click My Computer. If you can't find the My Computer icon anywhere, do the following:

    a. Right-click in the empty space on your desktop.

    b. Select Properties.

    c. A window opens. Click the Desktop tab.

    d. Near the bottom of the window, click the Customize Desktop button.

    e. Another window opens. On the General tab, beneath the words Desktop Icons, place a checkmark in the My Computer box.

    f. Click OK.

    g. You are returned to the previous screen. Click Apply.

    h. Click OK.

    i. The My Computer icon appears on your desktop. Right-click it.

2. Select Properties.

3. A window opens. Near the bottom of it, you should see the name of the company that made your CPU (usually it is Intel or AMD) as well as the CPU's speed listed in measurements of "MHz" (megahertz) or "GHz" (gigahertz)—for example, 500 MHz or 1.5 GHz (see Figure 1-3).

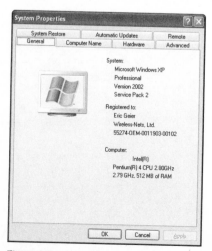

Figure 1-3

Open Your Computer's Case

After you have used Windows to do a quick analysis of your PC's primary features, you should open the computer's case and examine the equipment inside. Becoming comfortable with opening the case and identifying the components is essential if you ever want to perform a hardware upgrade on your own.

1. Turn off your computer and unplug it from all electrical outlets. This will prevent you from receiving a serious electrical shock and will protect the computer from being damaged.

2. Examine your case to find out how it opens. Some cases have a removable cover shaped like an upside-down "U" that slides off when some screws are unfastened. Other cases have side panels (or doors) that slide off or swing open when they are unscrewed or when a special button is pressed.

3. If you cannot figure out how to open your computer's case, check the owner's manual or documents that came with your system.

What Should Be Upgraded?

The top questions asked by most digital do-it-yourselfers are "When do I need to upgrade?" and "What parts should be upgraded?" Here are some common PC problems and the hardware or software upgrades that might fix them:

- **Problem:** The computer takes an unusually long time to start up, to shut down, or to open programs and folders. Also, the computer feels sluggish.

  Solution: Upgrade to a faster processor (CPU), or add more system memory (RAM).

- **Problem:** The computer frequently freezes or crashes, often resulting in error messages or blue screens displaying strange text.

  Solution: Update or reinstall Windows, or add more system memory (RAM).

- **Problem:** Messages warn that the computer is running low on disk space or doesn't have enough room to install a program.

  Solution: Upgrade to a larger hard drive, or add a second hard drive.

Use Newer Gadgets on Older PCs

To make a modern peripheral work on an outdated version of Windows such as 98 or ME, you need to install small pieces of software called "drivers" that tell Windows how to interact with the device. Usually all peripherals come with an installation CD-ROM containing the proper drivers, but if you have lost this CD or never received it (which often happens when you buy a used peripheral from a place like eBay), you will need to download the drivers.

1. Open your Web browser and visit the manufacturer's Web site.

2. Find the section of the site either labeled Support or Downloads, and search for information about drivers.

3. You probably will need to know the make and model of your product (so you may need to look on its exterior for that information).

4. Find the drivers that are designed to work with your version of Windows. Usually, the downloads will come in the form of an executable program. If so, download the executable to a folder on your computer that you can easily locate.

5. Double-click the executable to begin the installation of the drivers.

6. Try connecting your product again.

Solve Common Problems

Like any high-tech equipment, digital music players are not perfect. Here are some common problems and their five-minute fixes.

Music Player Can't Connect to Your PC

- Check the owner's manual to see if you are using the player correctly.
- Try connecting your player to a different computer.

Songs Won't Play Properly or Sound Distorted

- Your music files might be corrupt or infected. To avoid this problem, do not download illegal, pirated songs through file-trading software such as LimeWire, Kazaa, or iMesh.
- To guarantee that the songs you download are completely legal, free of distortion, and of high digital quality, buy them from official online music stores.

Battery Life Drains Quickly

- Do not listen to your music at an extremely loud volume. Instead, use a medium level.
- Do not listen to a snippet of one song and then immediately jump to a new one. Instead of forcing your player to constantly search its memory for the songs you want — which can drain battery power — try listening to an entire song before selecting a new one.
- If your player has a backlight, use it sparingly.
- If your player has an audio equalizer or bass booster, use it sparingly.

Windows Media Audio Files Won't Play Properly or at All

- Your Windows Media Audio (WMA) files might be using a copy protection feature such as DRM (digital rights management) or PRM (personal rights management). Usually these protection schemes are turned on when the WMA files are created. Therefore, if the songs were not originally created to be used solely on your computer, there isn't much you can do to make them play.
- If you use Windows Media Player or other CD-ripper programs to make WMA files from your music CDs, disable the option to use copy protection or DRM. Here's how to do it in Windows Media Player:

Note

This five-minute fix requires you to access Windows Media Player's "menu bar," which contains the drop-down menus you must click in order to change the player's settings. If the menu bar does not automatically appear at the top of your player, you can make it visible by simultaneously pressing the Ctrl key and the M key on your keyboard.

1. Open Windows Media Player.

2. Click the Tools drop-down menu.

3. Select Options.

4. A window opens. Click the Rip Music tab.

5. Look for the Copy protect music option. Next to this option, remove the check mark from the box.

6. Click Apply.

7. Click OK.

Laptop Battery Life Is Too Short

Laptop computers are quite convenient—assuming they have power. If you can't plug a laptop into an electrical outlet, it has to rely on batteries. Most laptops run only two to three hours on a fully charged battery. To extend the life of your laptop's battery, do the following:

1. Lower the brightness of your screen.

2. Disable your wireless card. To do so, double-click the wireless icon in the lower-right corner of Windows, and then click Disable (see Figure 7-3).

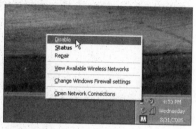

Figure 7-3

3. To reduce the battery's usage and extends its life, configure the power options in Windows.

For Windows XP and Windows 2000

a. Click the Start button in the lower-left corner of Windows.

b. Click Control Panel. If you don't see this option, your Start menu is in Classic mode. In that case, click Settings and then select the Control Panel.

c. If the Control Panel is in Category view, click the Performance and Maintenance category and then click the Power Options icon. If the Control Panel is in Classic view, simply double-click the Power Options icon.

d. Select an appropriate Power Scheme (see Figure 7-4).

e. Click OK.

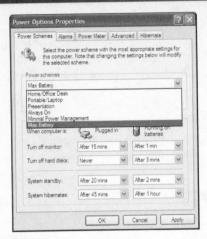

Figure 7-4

For Windows 98 and Windows ME

    a. Click the Start button in the lower-left corner of Windows.

    b. Click the Control Panel.

    c. Double-click Power Management.

    d. Select an appropriate Power Scheme.

    e. Click OK.

4. If your battery continues to last for less than one hour, replace it.

Protect Your Computer While You're Temporarily Away

If you need to step away from your computer for a few minutes, it could be vulnerable to intrusion by anyone who has physical access or Internet access to your system. Here is what to do to prevent this.

Log Off

The safest way to protect your computer while you are away is to log off. But be advised that logging off requires you to shut down all files, folders, and programs you are using, so save your work first.

For Windows XP Home Edition and XP Professional Edition

1. Click the Start button in the lower-left corner of Windows.

2. Click the Log Off button.

3. A window opens containing another Log Off button. Click it.

For Windows 2000

1. Click the Start button in the lower-left corner of Windows.

2. Click Shut Down.

3. In the drop-down menu, select Log Off.

Types of Phishing

E-mail links: The link sends you to a Web site that looks legitimate. When you follow the e-mail's instructions and verify or update your account information, your data is stolen and used to commit identity theft. Also, clicking the link can cause spyware or similar digital threats to be downloaded to your computer.

Instant-message links: Similar to e-mail phishing. These instant-message links are designed to look like they are from someone on your "buddy" list of contacts.

Pharming: This scam redirects you from legitimate Web sites to fake ones that look like the real deal. When you enter an ID, password, or credit card number, your information is stolen.

Cross-site scripting: Criminals sabotage real Web sites and put their own log-in boxes on those sites. When you enter an ID and password into the boxes, your information is stolen.

URL hijacking: Criminals take advantage of a company's flawed or unprotected Web site address to redirect you to a phishing site.

Phone or snail-mail scams: You receive a live phone call, a voicemail, or a snail-mail letter from a criminal who claims to represent a company or financial institution that you have done business with recently. The criminal asks you to verify or update some account information.

Canning Spam

If you want to can your spam problem and reclaim your Inbox from the onslaught of ads, get-rich-quick schemes, and X-rated messages, there are several steps to take.

Note

Even the best anti-spam strategies allow an occasional spam message to slip through. None of them is perfect by itself, but when used in combination with each other, they should reduce your spam problem by 95 to 99 percent.

Never open spam: There's an old saying about food products past their expiration date: "When it doubt, throw it out." This adage is true for real Spam as well as spam e-mail. If you suspect an e-mail is spam, just delete it. Some spam are actually designed to alert their creators whenever they are opened, which lets the spammer know that your e-mail address is active and ripe for more spam. If you are tempted to open questionable e-mail because you think it is safe, you can view details about it by using the "passive viewer" of Outlook or Outlook Express. For more information, refer to Chapters 3 and 4.

Never click spam links: If you accidentally open a spam, close it immediately. And no matter what, never click any links in the spam. Doing so could cause you to become a victim of a "phishing" scam.

Turn on your ISP's spam filter: Not all Internet service providers (ISPs) turn on spam filters by default when you sign up with them or when you add another e-mail address to your existing account. Check with your ISP to ensure that all possible spam filters are being used.

Install anti-spam software: Numerous companies make special software that will plug into your favorite e-mail program to scan your incoming messages and block almost all spam from entering your Inbox. Here are some popular programs (in no particular order):

- ETrust Anti-Spam (www.ca.com)
- Spam Shredder (www.webroot.com)
- Norton AntiSpam (www.symantec.com)
- iHateSpam (www.sunbelt-software.com)
- McAfee SpamKiller (www.mcafee.com)
- MailWasher Pro (www.mailwasher.net)
- CYBERsitter Antispam (www.cybersitter.com)
- Cloudmark SafetyBar (www.cloudmark.com), free for Outlook and Outlook Express

Use multiple e-mail accounts: To increase your privacy and reduce spam, use multiple e-mail accounts. Keep one "good" account for e-mailing friends and family, and have a "junk" account used for Internet purchases and for documents requiring an e-mail address (from banks, credit cards, club memberships, doctors' offices, and so on). Doing so keeps your "good" address from being sold to third-party marketers. When signing up for a "junk" account, divulge as little private information as possible. That way, if your "junk" address gets passed around the Internet, your confidentiality will be protected.

Take precaution with Internet messages: If you post messages on Internet newsgroups (often referred to as the "UseNet") or on "blogs," do not include your real e-mail address with your message. Sinister programs called "spambots" scour newsgroups and blogs daily to harvest e-mail addresses to which marketers can send spam. A solution is to post messages that contain an altered version of your e-mail address along with instructions on how to decipher it. For example, you could write `Reply to: bob@nospamhotmail.com` and remove the `"nospam"`.

Disable the Preview pane: In Microsoft's Outlook and Outlook Express, a split-window feature known as the Preview pane can help spammers to send you more junk mail.

Block Pop-Ups

Make your Internet experience faster, safer, and less cluttered by using a pop-up blocker. This simple tool prevents your computer from being bombarded with endless, annoying Web advertisements. This is one of the easiest things you can do to reduce online hassles — and it won't cost you a cent!

Built-in blockers: The latest releases of popular Web browsers such as Internet Explorer, Firefox, and Opera have pop-up blockers built into them. If your browser is outdated, this is a great reason to upgrade.

Note

To use Internet Explorer's pop-up blocker, you must install Service Pack 2 from the Windows Update Web site. For more information, refer to Chapter 1.

Toolbars: Several reputable companies offer free toolbars that hook into your Web browser to block pop-ups and provide Internet search capabilities.

- Yahoo Toolbar (`http://toolbar.yahoo.com`); also has an anti-spyware feature known as Anti-Spy

- MSN Toolbar (`http://toolbar.msn.com`)

- Google Toolbar (`http://toolbar.google.com`)

Delete Data Once and for All

- The only safe way to get rid of your files is to wipe them from existence — literally. Special software exists that can *wipe* data (also known as *shredding*), which will make the files almost impossible to recover.

- During the wiping process, your old files are overwritten numerous times by new, random data. Think of it like painting the same wall in your house over and over with a different color each time.

Close an Unresponsive Program

Occasionally, a program may throw the digital equivalent of a temper tantrum and refuse to close. When that happens, you can force it to shut down by using the Task Manager, as follows:

1. Simultaneously press the Ctrl, Alt, and Delete keys on your keyboard, which opens the Windows Task Manager. (However, if your version of Windows is configured differently, ressing these keys might open a Windows Security box. In that case, simply click the Task Manager button.)

2. Click the Applications tab.

3. Click the name of the unresponsive program.

4. At the bottom of the Task Manager, click the End Task button.

5. If the troubled program doesn't close immediately, a message alerts you that the program is not responding. Click the End Now button.

6. If the program still does not respond, or if Windows feels sluggish, shut down your computer and restart it.

Disable Automatic Cleanup of Your Desktop

To keep your desktop free of clutter, Windows XP offers to automatically relocate any shortcuts that haven't been used in quite a while. If you are satisfied with the layout of your desktop and don't want it changed, you should disable the Desktop Cleanup Wizard, like this:

1. Right-click in the empty space on your desktop.

2. Select Properties.

3. A window opens. Click the Desktop tab.

4. Click the Customize Desktop button located in the lower-left corner.

5. Remove the check mark from the Run Desktop Cleanup Wizard Every 60 Days box.

6. Click the OK button.

Easily Locate Your Downloads

Have you ever downloaded a program, video, or picture from the Internet but been unable to find where it went? No matter where you search, you just can't seem to locate it. And because you can't remember the name of the download, the Windows Search Companion is useless. This is a common problem for many Internet surfers. A quick, easy solution is to create a special folder on your hard drive that will store all of your downloads. Here's how:

1. Double-click the My Computer icon on your desktop. If this icon is not available, click the Start button in the lower-left corner of Windows and click My Computer. If you can't find the My Computer icon anywhere, do the following:

   a. Right-click in the empty space on your desktop.

   b. Select Properties.

   c. A window opens. Click the Desktop tab.

   d. Near the bottom of the window, click the Customize Desktop button.

   e. Another window opens. On the General tab, beneath Desktop Icons, place a check mark in the My Computer box.

   f. Click the OK button.

   g. You are returned to the previous screen. Click the Apply button.

   h. Click the OK button.

   i. The My Computer icon appears on your desktop. Double-click it.

2. A window opens. Double-click the icon for your C: drive (unless you installed Windows in a different location, in which case, double-click that drive letter).

3. Click the File drop-down menu.

4. Select New.

5. Select Folder.

6. A new folder (aptly named New Folder) appears in your C: drive. Right-click it, and then select Rename.

7. Type a new name for the folder such as Downloads or My Downloads.

8. Each time you download a new file or program, save it to your new downloads folder so that you always know where to find it.

Protect Your Computer from Viruses

To prevent infection from digital viruses, create some good habits by following all of these tips:

- Install trusted, respected antivirus software and keep it updated constantly.

- Do not open e-mail attachments that have a file extension of .exe, .scr, .vbs, or double file extensions like .txt.vbs.

- Be wary of opening any e-mail attachments or instant-message attachments sent from people you don't know — even if those attachments do not have a dangerous file extension.

- Do not open spam e-mail (selling products, offering free videos, pictures, or songs, and so on).

- Perform a virus scan on files before downloading or opening them.

- Perform a virus scan on e-mail attachments you think are safe to open.

- Do not install pirated software, because often it contains viruses.

- Do not download pirated music files or videos, because they too contain viruses.

- Do not click links sent to you in an instant message.

Rearrange Your Desktop Shortcuts

Most computers running Windows XP automatically use a feature known as Align to Grid that attempts to bring order to your desktop by stacking your shortcuts in clean, evenly-spaced rows and columns. Unfortunately, this feature prevents you from customizing the look of your desktop. Many people enjoy the ability to freely move their shortcuts anywhere because it makes their desktops feel less cluttered and allows more of their desktop wallpaper to be seen. To arrange your shortcuts how ever you want, turn off the Align to Grid option as follows:

1. Right-click the empty space on your desktop.

2. Select Arrange Icons By.

3. Select Align to Grid.

4. Now you can arrange your shortcuts anyway you want by clicking them and dragging them to any spot on your desktop. To enhance the beauty of your desktop wallpaper and show as much of it as possible, consider placing your icons along the edges of the desktop.

Rename Several Files at Once

If you want to give new names to a group of files that share a common theme—
such as a collection of digital photos taken during a vacation—you can save
yourself time and finger cramps by renaming them all at once (known as a
"batch rename"). Follow these steps:

Note

A batch rename can be done in only one folder at a time. Also, be careful not to
accidentally rename any critical system files. Doing so could cause Windows to mal-
function.

1. Select the files you want to rename using one of the following methods:

   • Simultaneously press the Ctrl key and the A key to select all of
     the files in a folder.

   • Click the first item in a list, and then hold down the Shift key and
     click the last item in the list. This causes the first and last items
     and any between them to be highlighted.

   • Hold down the Ctrl key on your keyboard while clicking each
     individual item you want to select.

2. Right-click the file you want to go first in the newly renamed series.

3. Select Rename.

4. Type a name for the first file in the series, and then press the Enter key.

5. All of your selected files are given this new name, followed by a number that distinguishes them from one another. For example, if you
   renamed the first file Geeks On Call, the rest of the files would be
   named Geeks On Call 1, Geeks On Call 2, Geeks On Call 3, and so
   on.

6. If you are unhappy with the new names, you can restore the original
   ones by simultaneously pressing the Ctrl and Z keys on your keyboard. Each time you use this keyboard shortcut, only one file reverts
   to its original name. That means you must use this shortcut numerous
   times to restore all of the names.

Make a Screen Capture

On occasion, you may wish to take a snapshot of your Windows desktop or the files inside a folder or window. Here's how:

1. Do one of the following:

- To capture an image of your entire screen, press the Prt Scr key.

- To capture an image of a particular window or folder, open it, and then simultaneously press the Alt key and the Prt Scr key.

2. After the image is captured, you can edit or print it by pasting it into Microsoft's Paint program as follows:

  a. Click the Start button in the lower-left corner of Windows.

  b. Click on All Programs.

  c. Select Accessories.

  d. Select Paint.

  e. After Paint opens, click the Edit drop-down menu.

  f. Select Paste.

  g. To print the screen capture, click the File drop-down menu.

  h. Select Print.

3. Another option is to paste the screen capture into a photo-editing program like Adobe's Photoshop or Photoshop Elements, Microsoft's Digital Image or Picture It, or Ulead's PhotoImpact or Photo Express.